ARABLE PLANTS
– a field guide

Phil Wilson

Miles King

ENGLISH NATURE   **WILD**Guides

First published 2003 by **WILD**Guides Ltd.

WILDGuides Ltd.
Parr House
63 Hatch Lane
Old Basing
Hampshire
RG24 7EB

www.wildguides.co.uk

ISBN 1-903657-02-4

This book is accessible in electronic format via the following URL:

http://www.english-nature.org.uk/science/botany/default.htm

Production and design by WILD*Guides* Ltd., Old Basing, Hampshire.
Printed in China. The Hanway Press, London.

Front Cover photograph: Poppies in Linseed Field: Paul Glendell/English Nature.

Contents

SPECIES ACCOUNTS

BROAD-LEAVED SPECIES

English Nature

Places where wildlife live have been under great pressure in recent decades. Many species of plants and animals have been declining and loss of habitat amounts to an area the size of the county of Shropshire every ten years. English Nature works to champion wildlife and to redress the balance by:

▸ overseeing a system of protected sites and wildlife legislation;
▸ introducing suitable management practices;
▸ devising and implementing wildlife-friendly policies;
▸ working in partnership with a range of people and organisations; and
▸ providing scientifically-based, sustainable solutions.

Some of the improvements are reflected in agri-environment schemes; reductions in the use of pesticides, increases in populations and the range of both Red Kites and Otters, legal protection for species including the Water Vole and Great Crested Newt, the re-introduction of the Large Blue Butterfly declared extinct in the UK in 1979 and an increased awareness of the importance of plants by safeguarding them and the sites where they grow.

We want people to understand and appreciate the importance of England's natural heritage and produce a range of information which is available from: English Nature. Their headquarter address is:

Northminster House, Peterborough PE1 1UA;
Tel: 01733 455000; Fax: 01733 568834.

There are also 22 Local Team offices, details of which can be obtained by telephoning Northminster House or by obtaining a copy of *English Nature Facts and Figures* information guide free from the Enquiry Service **Tel: 01733 455100**.

You can also learn more about us via the internet. Our address is:

www.english-nature.org.uk

WILD*Guides*

WILD*Guides* is a publishing organisation with a commitment to supporting wildlife conservation through financial donations and the provision of professional services. Whilst WILD*Guides* specialises in the design and publication of definitive yet simple-to-use identification field guides to wildlife, as an organisation it is also fully equipped as a creative agency.

WILD*Guides* can be contacted at:

Parr House, 63 Hatch Lane, Old Basing, Hampshire RG24 7EB;
Tel: 07818 403678; Fax: 01256 818039.

www.wildguides.co.uk

Foreword

Where have our cornfield flowers gone? Fields red, blue and yellow with Common Poppies, Cornflowers and Corn Marigolds are a distant memory. Those rare occasions where a crimson splash of poppy flowers show where a herbicide spray has been ineffective invariably excite comment.

Many of our once common arable plants are now among our rarest wild flowers. This marked decline is such that 12 of the 66 plant species listed on the government Biodiversity Action Plan are arable plants and the cereal field margins where they largely occur are also included amongst the habitats requiring urgent attention. It is vital that agri-environment schemes take these species into account. This situation is not just confined to Britain, but is repeated in every other European country in response to the massive and far-reaching changes in arable farming that have occurred since the middle of the 20th century. One of the ironies, of course, is that farming has been designed to eliminate these plants and, as crops and management have altered, so the variety of the arable flora has declined.

The aim of this book is to act as a spur to the interest of all conservationists, farmers, agricultural advisers and others concerned with the biodiversity of Britain's farming landscape. This field guide draws together the available information about Britain's arable flora to help land managers and advisers in its conservation. Above all, it is hoped that it will raise awareness of these fascinating and beautiful plants and help their conservation.

SIR MARTIN DOUGHTY
CHAIR
ENGLISH NATURE

Neolithic

Introduction

Arable farming has always been a struggle against the forces of nature. Early farmers had to deal with difficult climates, unfriendly soils and often steep slopes. Perhaps the greatest challenge though was the eternal battle with the many 'weeds' of the crop. These flowers shared the same ecological niche as the crop plants with which they cohabited: they thrived on regular disturbance and a short growing season.

A weed or not a weed?

A few weeds still cause major crop losses, and this might lead some arable farmers to wonder whether it is worth conserving any wild plants on arable land. Until very recently, only a few botanists were worried about the disappearance of these plants. But this has changed now that familiar plants such as Cornflower and Corn Marigold are now very rare, and the Corncockle is virtually extinct in the wild. Arable land also has a reputation for being a desert for wildlife, though well-managed arable farmland can be exceptionally rich in birds, mammals, insects and plants. This book will show how simple it is for arable flowers and modern arable agriculture to live happily, side by side.

The arable context

Arable land currently occupies approximately seven million hectares (nearly 30%) of Britain's land surface. Much of the arable land lies on the eastern side of Britain where it is warmer and drier, allowing it to be ploughed more easily. Permanent pasture is concentrated to the west of Britain and in Ireland.

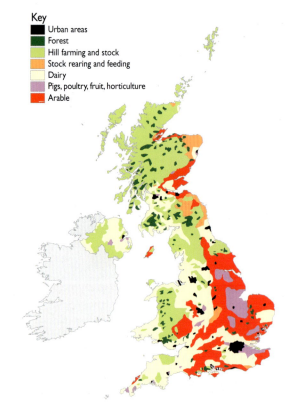

Key
- Urban areas
- Forest
- Hill farming and stock
- Stock rearing and feeding
- Dairy
- Pigs, poultry, fruit, horticulture
- Arable

Types of farming in Britain
(based on map p.5, *Britain Today: Farming*, Woodcock, R., Wayland Books, 1994).

Farming past

Arable farming has been a feature of the British landscape for some 8,000 years. Around 6000 BC, the hunter/rancher lifestyle, prevalent since the Ice Age, was gradually replaced by a culture, originating in the Middle East, which depended upon domesticated animals and cultivated crops.

Crossing continents

The wild plants that benefited from agriculture originated from a range of different habitats. Many were present in Britain before the introduction of arable farming. It is important to realise that, just like plant communities of grasslands and woodlands, arable plant communities have distinct relationships to soil, climate and management. They have not just been thrown together in recent centuries. Britain's arable plants are particularly important, as they occupy the north-western end of a range of communities spreading across Europe into Asia; most have declined over their whole European range during the last century.

Well-established

Arable plants are not the fly-by-night colonists that some might think. Some populations have been known from particular sites for many years. Even though individual plants are annual, arable plant communities are surprisingly stable. There is evidence that fields with a long history of arable cultivation have the richest communities of arable plants. Arable plants are best described as native to a particular land-use rather than a geographical region.

Cultural links

Throughout Europe, the flowers of arable crops have a special place in the public psyche. This connection goes back many centuries to a time when peasants worked with and against these plants every day. Even those arable plants that are now very rare must once have been familiar cornfield inhabitants.

Broad-leaved Cudweed (*top*), Ground-pine (*middle*) and Rough Marsh-mallow have been known from one field in Kent since the end of the 18th century. Corn Buttercup, Shepherd's-needle (*bottom*), Corn Cleavers and Spreading Hedge-parsley have occurred on the Broadbalk experimental plot at the Rothamsted Experimental Station since its inception in 1843.

Corncockle (*left*) and Darnel were among the most serious 16th century weeds, and feature in the works of William Shakespeare and John Donne. Even now, Cornflower is readily understood as a particular shade of blue. The use of arable flowers as literary metaphors shows how significant they were in former times. Few people now would know them as wild plants.

Flanders Field

The poppy has entered the cultural landscape perhaps more than any other arable plant and is still readily recognised throughout northern Europe. It has as great a cultural resonance as any other European plant, and has been a symbol of rebirth and new life since ancient Egypt. This symbolism entered new dimensions in the aftermath of the appalling destruction of the battles of Ypres and the Somme between 1914 and 1917. The battlefields bloomed with sheets of blood-red poppies (and several other arable species) in the summers following these battles, and have entered the literature, mythology and the traditions of a whole continent. In France, the Cornflower occupies a similar symbolic position. Paper poppies are still worn in Remembrance in Britain. Poppies also feature in works by the artists Claude Monet and August Renoir, amongst others.

Garden flowers

The poppy has also found its way into gardens in a variety of forms following many years of breeding. Other arable plants are also cultivated for ornamental purposes including Cornflower, Thorow-wax and Corncockle. Pheasant's-eye has been used as a cut-flower, with large quantities being gathered from the Sussex Downs at the end of the 18th century and sold at Covent Garden when it was known as 'Red Morocco'.

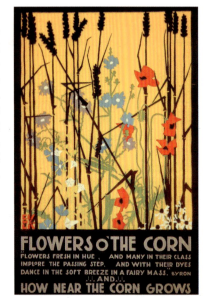

Artists such as E. McKnight Kauffer were commissioned to produce posters. This one – *Flowers o' the Corn* – illustrates that in 1920 cornfield flowers were still widespread compared with today.
With kind permission of London's Transport Museum.

FLOWERS O'THE CORN
FLOWERS FRESH IN HUE , AND MANY IN THEIR CLASS
IMPLORE THE PASSING STEP, AND WITH THEIR DYES
DANCE IN THE SOFT BREEZE IN A FAIRY MASS." BYRON
...AND...
HOW NEAR THE CORN GROWS

Going, going …

The arable flora has undergone great changes throughout Europe since the end of the 19th century, and the losses accelerated towards the end of the 20th century. Most arable plants were overwhelmed by the massive revolution in arable farming methods, including:

‣ more efficient seed-cleaning techniques;
‣ the widespread adoption of herbicides;
‣ the development of highly nitrogen-responsive crops;
‣ the increase in nitrogen applications;
‣ the near-complete mechanisation of farming;
‣ changes in crop rotations; and
‣ efficient field drainage.

As a result, arable land in Britain has lost most of its arable plants: several species have become extinct and many more are now rare. These include some that were once extremely common and caused serious farming problems. Cornflower is now one of Britain's most endangered plants, but until the mid-19th century it was abundant, *"a pernicious weed injurious to the corn and blunting the reapers' sickles"*. Corncockle and Darnel are now virtually extinct in Britain, although both remained locally common until the early 20th century.

Technology that allowed better seed-cleaning caused an initial decline in arable plants in the late 19th century, but herbicide development in the 1940s was catastrophic for many more. Corn Buttercup and Shepherd's-needle were both abundant until the early 1950s; indeed, both are listed in early weed-control handbooks with recommendations for their control. Hefty increases in nitrogen application and the development of highly competitive crop varieties placed additional pressure on many arable plants and may have been the major factor in the extinction of Small Bur-parsley and Lamb's-succory.

The plants of irregularly cultivated field edges, and other places which are disturbed from time to time, have also suffered from increasingly efficient farming methods. Their plight may be even greater than that of more conventional arable plants, as their special requirements are less well understood. These plants cannot compete under the shade of an arable crop and have life-cycles that do not fit well with the timing of farming practices. Many have always been restricted to south-facing chalky slopes in the warmer parts of southern England, and include such rarities as Ground-pine and Cut-leaved Germander. The dry, sandy soils of the East Anglian Breckland are also good for these plants.

Uniform fields, uniform weeds

Arable land has become increasingly uniform, due to:

‣ the continual use of one type of crop;
‣ the use of herbicides; and
‣ massive increases in the applications of nitrogen.

Climatic, soil and management factors, which once led to diversity of arable habitats, are now much less important. The formerly common suite of arable plants has all but gone, replaced by a small but pernicious gang including Black-grass, Cleavers and Barren Brome. These have thrived under intensive management and now occur over large areas of Britain.

Arable flowers have declined across a range of scales: at the country level, the farm and even the field scale. On conventional farms, almost all uncommon arable plants are confined to the 4 m strip along the field edge. This is due to the irregular application of herbicide and fertiliser, the less efficient drilling of the crop and the effects of soil compaction. So, even where arable plants have survived, they are confined to tiny areas.

This restriction to field margin refuges renders them vulnerable to yet another threat. Hedgerows and other field boundaries have been removed on a massive scale in Britain and the rest of Europe to facilitate the use of large machinery. Only half the length of hedgerow present in Britain in 1945 was still present in 1990. Not only has this removed the physical boundaries, but also the strip of less-intensively farmed land alongside the boundary where the more diverse flora can survive.

Survival

Despite all the pressures under which the arable flora has suffered, there are some areas of Britain where uncommon arable plants still persist.

Farmland birds

Other farmland wildlife has also suffered in recent years. Birds have been particularly well-studied, and of the 26 species listed as priorities in the UK Biodiversity Action Plan (UKBAP), 13 are predominantly species of lowland farmland. These include the Grey Partridge, Turtle Dove, Tree Sparrow, Skylark and Corn Bunting, all of which have declined catastrophically.

The major reasons that have been identified for these declines on arable land are:

▸ reduction in the area of spring-sown cereals, and the loss of crop stubbles in the winter and bare ground in the spring;

▸ loss of mixed farming due to simplification of crop rotations and regionalised specialisation;

▸ increased use of pesticides and fertilisers; and

▸ loss of field boundaries.

These have removed nesting sites and reduced availability of food during the breeding season (e.g. food for chicks) and over the winter.

Corn Bunting (*above*) and Skylark (*below*).

The majority have retreated from the north of the country, and few can now be found north of Yorkshire. A few, like Corn Marigold, are still relatively widespread particularly in western Scotland.

Hot spots

The richest areas are in the south-east of England, particularly where soils are light and chalky. Fields around the coasts of south-west England and Wales can, however, also be very rich in arable plants of both calcareous and acidic sandy soils: a National Trust property in north Cornwall is one of Britain's few arable plant reserves. The heavy calcareous soils of the mid-Somerset hills have several sites for uncommon arable plants like Spreading Hedge-parsley and Broad-leaved Spurge, and one of these is managed as a Somerset Wildlife Trust reserve. There are outstanding areas for arable plants on the chalk between Salisbury and Basingstoke, and in south Cambridgeshire. The Breckland

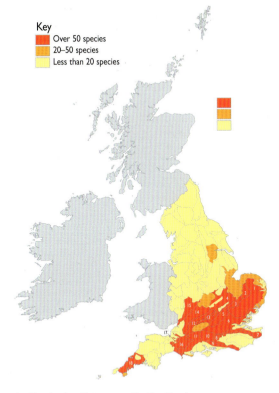

Key
- Over 50 species
- 20–50 species
- Less than 20 species

1	East Anglian Plain	2	Breckland
3	East Anglian Chalk	4	West Anglian Plain
5	Chilterns	6	London Basin
7	North Downs	8	Wealden Greensand
9	South Downs	10	Hampshire Downs
11	South Wessex Downs	12	Thames and Avon Vales
13	Mid Vale Ridge	14	Cotswolds
15	Severn and Avon Vales	16	Wessex Vales
17	Mid Somerset Hills	18	Cornish Killas and Granites

The best areas in England for arable plants.

of Suffolk and Norfolk is unique in Britain with its sandy soils and continental climate; it is home to an extraordinary variety of arable plants, several of which are found nowhere else in Britain.

National and international policy

The politics of agricultural production has had a direct effect on the fortunes of arable plants. In recent years, the emphasis on economic planning after the Second World War resulted in the 1947 Agriculture Act. This was superseded on Britain's accession to the Common Market by the Common

Agricultural Policy (CAP). The overall effects of the Agriculture Act and the CAP have been to encourage the intensification of arable farming by subsidising production and guaranteeing markets for surpluses. This resulted in the rapid development of agrochemicals, artificial fertilisers and farm mechanisation, and led directly to the removal of hedgerows and field boundaries in the name of efficiency – all to the detriment of Europe's arable flora and other arable wildlife.

Attempts to redress the effects of intensification began in Britain in the late 1980s with the introduction of pilot agri-environment schemes. These measures were subsequently subsumed into the government-run Environmentally Sensitive Area and Countryside Stewardship Schemes. The prospects for arable plants now look even better following the recent introduction of Arable Stewardship. Under this option of the Countryside Stewardship Scheme, farmers can be paid to manage their land to encourage the development of arable plant communities.

Further attention has been directed towards the biodiversity of arable land as a result of the UK Biodiversity Action Plan (UKBAP). This was published in 1995 as the UK government's response to the Biodiversity Convention, which was signed at the Earth Summit in Rio de Janeiro in 1992. Twelve arable flowering plants and three mosses are included on the list of species of priority concern in the UKBAP (Appendix 3). In addition, cereal field margins are included as a habitat in need of urgent conservation action. A nationally-costed Action Plan is now in place for each of the species and for cereal field margins.

A typical weedy field margin.

Bronze Age

Origins and development

The deliberate cultivation of cereals began more than 12,000 years ago. Wild forms of Einkorn and Emmer, both primitive forms of cultivated wheat, Two-row Barley and several legumes were originally domesticated in eastern Turkey and northern Iraq, and would have been accompanied by annual weeds from the start. Traces of settled arable farming have been found in Greece from approximately 8,000 years ago, as the Neolithic culture crept gradually westwards, displacing the former hunting/ranching systems.

First British farmers

Arable agriculture moved more rapidly across the rest of Europe, arriving in Britain approximately 7,500 years ago and spreading across most of the country in the following 1,500 years. The earliest fossil record of cereal pollen comes from the Scottish island of Arran about 8,000 years ago. Arable farming was initially concentrated on the chalk and limestone plateaux of the Wessex Downs and the Cotswolds. These early farming communities were centred on large earth settlements, the remains of which can still be seen on many hill-tops in Dorset and Wiltshire. Here, the loamy, calcareous soils were sufficiently deep and fertile to support regular cropping, but were naturally well-drained and relatively easily cleared. The major cereals were Emmer Wheat and Six-row Barley, together with Einkorn Wheat, Flax, Bread Wheat and Spelt Wheat.

Einkorn (*above*) and Emmer (*below*).

Major Neolithic weeds:
Common Poppy, Common Fumitory, Charlock, Wild Radish, Penny-cress, Corn Spurrey, Four-seeded Vetch (Smooth Tare), Pale Persicaria, Small Nettle, Ivy-leaved Speedwell and Field Madder.

Butser Ancient Farm, Hants: a replica of a British Iron Age farm, c. 300 BC.

Crops and weeds develop

The archaeological record gives us some clues to the problems that Neolithic farmers may have faced (Appendix 2). Arable plants that are still common today were dominant in the earliest arable flora, and species which are now rare, including Rye-brome, Cornflower, Narrow-fruited Cornsalad and False Cleavers, have all been found in Neolithic arable deposits.

Origins

Where did these plants come from? Many of Britain's arable plants have been found by archaeologists to pre-date the start of arable farming, so it is clear that these species were here before the first plough touched the soil. Some still occur in other habitats in Britain, and may have originated here. For example:

▶ coastal shingle and grassland support rare arable plants such as Red Hemp-nettle, Small-flowered Buttercup and Corn Parsley;
▶ grassland and woodland support common arable plants, including Common Hemp-nettle, Chickweed, Shepherd's-purse, Scarlet Pimpernel and Parsley-piert in disturbed habitats; and
▶ disturbed ground supports some rare arable plants like Ground-pine and the cudweeds.

Other species may have been introduced with grain brought from elsewhere in Europe, but the origins of these are lost in the mists of time. Some can no longer be found anywhere in any habitat other than arable land, and

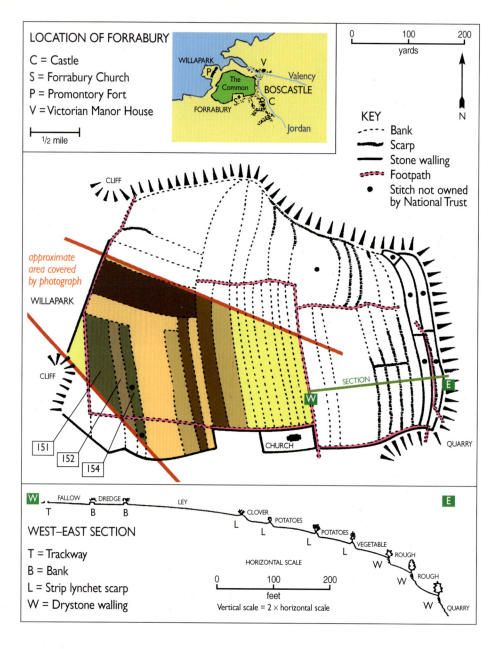

LOCATION OF FORRABURY

C = Castle
S = Forrabury Church
P = Promontory Fort
V = Victorian Manor House

1/2 mile

WILLAPARK
P
The Common
FORRABURY
S
V
Valency
BOSCASTLE
C
Jordan

0   100   200
yards

KEY

- - - - Bank
∼∼∼∼ Scarp
──── Stone walling
▬▬▬▬ Footpath
● Stitch not owned by National Trust

N

CLIFF

approximate area covered by photograph

WILLAPARK

CLIFF

151
152
154

SECTION
W
E
CHURCH
QUARRY

WEST–EAST SECTION

T = Trackway
B = Bank
L = Strip lynchet scarp
W = Drystone walling

W   FALLOW   DREDGE   LEY   E
T   B   B   CLOVER
L   POTATOES
L   POTATOES
L   VEGETABLE
W   ROUGH
W   ROUGH
W   QUARRY

HORIZONTAL SCALE

0   100   200
feet
Vertical scale = 2 × horizontal scale

(Above) The Forrabury Stitches open field system near Boscastle, west Cornwall. Plan and cross-section. Based on a map drawn by P. D. Wood in *Cornish Archaeology* No.2 (1963).

(Opposite top) Annotated photograph of The Forrabury Stitches showing the fields as highlighted on the plan.

24

151
152
154

most of these are now endangered throughout their ranges. In Britain, therefore, we must consider the conservation of the arable flora in an international context. It is more appropriate to treat these plants as natives of this particular type of land use rather than as indigenous to any part of Europe or the Middle East. The term 'archaeophyte' has recently come into use to describe species, such as these, whose origins are obscure, but which are thought to be ancient associates of human activities. This ancient association between humankind and the arable flora has earned them a special place in the way that we value them.

The Bronze Age, the Iron Age and the Romans

As more land came under cultivation between the Bronze Age and Roman times, so the numbers of 'weed' species increased. New crop species were introduced, including Hemp, Flax and Rye during the Iron Age,

Some surviving examples of open-field farming.

Laxton, Nottinghamshire
Partial survival of a working mediaeval system with three fields still farmed in strips by the commoners.

Forrabury, north Cornwall
Forty open-field strips separated by grass baulks survive on a cliff-top near Boscastle. Probable relics of the mediaeval open-field system in a county where such systems were rare.

Braunton, north Devon
The Great Field is still divided into strips which form part of the mediaeval open-field system.

Portland, Dorset
The small coastal fields are survivors of the mediaeval open-field strips divided by grass baulks. Until recently these were farmed by the inmates of the local prison, and had a rich arable flora.

Soham, Cambridgeshire
The North Field of the village is unenclosed and retains its strips. These are now owned privately.

and, in the Roman period, Opium Poppy, Lentil, Broad Bean, Pea and other vegetables. These crops would all have favoured different plants, and in recent times Rye and Flax were still known as particularly weedy crops. Each new importation of crop seed from mainland Europe would have also introduced its own seed contaminants. Corncockle, Rye Brome, Cornflower and the specialised weeds of Flax crops were transported in crop seed and re-sown along with it. During the Iron Age and Romano-British periods, arable farming was carried out in small fields arranged around single farmsteads or collections of dwellings. Such farms still survive in the West Penwith area of Cornwall, where the massive field walls date from the Bronze Age.

Food for free
Some species that are now regarded as weeds were deliberately cultivated for food in the past. Redshank, Fat-hen, knotgrasses, Rye Brome and Wild Oat were harvested for their large and nutritious seeds which were then made into unleavened bread (oats are thought to have originally been a weed of other arable crops). Such bread has been found with Roman remains in Britain. The last meal of the famous Iron Age 'Tollund Man', found in a Danish bog, consisted of Barley, Linseed, Black-bindweed, Pale Persicaria, Corn Spurrey and Fat-hen.

The Saxons and Feudalism
During the Saxon period large changes took place over much of the country. The chaos of the 'dark ages' led to the abandonment of much arable land. In the 8th and 9th centuries the Saxon open-field system was introduced to many parts of the country, including much of central southern England, the Midlands and eastern England from Norfolk northwards. Typically, each open-field system consisted of three large fields grouped around a village. These fields were divided into strips, each of which was allocated to a commoner, so that each commoner would have farmed several strips scattered throughout the fields. The same crop was grown by each farmer in the same field, and there would have been a simple crop rotation, so that one field each year would have had spring-sown Barley, one winter-sown Wheat and the third fallow.

Mediaeval period
We know more about the arable floras of mediaeval Britain. Archaeological remains are more abundant and better preserved, and there is documentary evidence for the plant problems that farmers had to cope with. Crops changed between the Roman period and the 12th century AD. The Spelt and Emmer Wheat of the early farmers were replaced by Club Wheat and Bread Wheat.

Rye and oats became very commonly grown and Barley was still frequent. Flax and Hemp were frequent fibre crops, but Beans, Peas and other vegetable crops were probably grown in garden plots rather than as field crops. Soil type, climate and farming methods would have caused differences between plant communities as they do today.

By mediaeval times, the arable plant communities of today were already well established. The most frequent weeds in mediaeval Britain were Rye Brome, Scented Mayweed, Wild Radish, Corncockle and various vetches and cleavers.

The earliest legislation against a weed was passed by Henry II in the 12th century and required the destruction of Corn Marigold. This species is also listed as one of the most troublesome weeds of agriculture, in Fitzherbert's Boke of Husbandry of 1523.

Published references to weeds became more common from the 17th century onwards as agriculture became more sophisticated. By 1700, agriculture had changed little from that of pre-Norman times. There was little understanding of plant nutrition, and crop rotations often consisted of just an alternation of cereal and fallow. More than half of Britain's arable land was farmed under an open field system. Crop yields were low and little effort was made to control weeds.

(Above) Bayleaf Farm, Weald and Downland Museum, Sussex.

The worst weeds of 1523:

thistles, docks, nettles, Corncockle, Darnel, Dodder, Corn Marigold, Hairy Tare, Scented Mayweed and Charlock.

Shakespeare's weeds

KING LEAR: ACT IV, SCENE IV

Cordelia:
Alack! 'tis he: why, he was met even now
As mad as the vex'd sea; singing aloud;
Crown'd with rank fumiter and furrow weeds,
With burdocks, hemlock, nettles, cuckoo-flowers,
Darnel, and all the idle weeds that grow
In our sustaining corn. — A century send forth;
Search every acre in the high-grown field,
And bring him to our eye.

CORIOLANUS: ACT 3, SCENE I.

Coriolanus:
In soothing them we nourish 'gainst our senate
The cockle of rebellion, insolence, sedition,
Which we ourselves have plough'd for, sow'd and scatter'd,
By mingling them with us, the honour'd numbers;
Who lack'd not virtue, no, nor power but that
Which they have given to beggars.

HENRY VI PART I

La Pucelle:
Good-morrow, gallants! want ye corn for bread?
'Twas full of darnel;— do you like the taste?

Roman

Agricultural Revolutions – the 18th–20th Centuries

The story of Britain's arable flora has been one of constant flux in response to changes in agricultural practices, with the pace of change accelerating rapidly from the beginning of the 18th century. William Pitt reviewed the agriculture of Leicestershire in 1809. Few of the species that he regarded as problems then would be seen as such today. Present-day rarities that he included as less important weeds were Shepherd's-needle, Corn Cleavers, Cornflower and Corncockle. The now rare Pheasant's-eye and its frequent associate, Thorow-wax, now extinct in Britain, were locally common in the south of England at the beginning of the 19th century. Flax crops had their own distinctive associates including Gold-of-pleasure, Flax Dodder and Flax Catchfly, all of which have now virtually disappeared from Europe. What were the developments that caused these transformations?

The first agricultural revolution – improving the land
The beginning of the 18th century there were significant improvement in agricultural practice. Jethro Tull developed the horse-drawn seed-drill, and four-course rotations including clover and roots were introduced under the influence of Lord 'Turnip' Townshend. These changes would have been impossible under the old common-field system, but for a succession of Enclosure Acts which enabled landowners to evict the peasantry from the land to work in the rapidly growing industrial cities. The amalgamation of narrow strips into large fields managed by a single farmer also enabled the introduction of horse-drawn hoes, harrows and much improved ploughs. This new technology was by no means adopted rapidly, and traditional methods were still widely practised well into the 19th and even the 20th centuries.

Under the Corn Laws of 1773 and 1815, the government paid a bonus on exported crops but taxed imported grain. This encouraged the expansion of arable land and the intensification of production. These laws made the enclosure of common fields and the eviction of commoners even more advantageous to landowners, and were much resented. Although the laws were repealed in 1846 to make way for free-trade in cereals, their effects persisted until the 1880s. The associated peak in the arable acreage of Britain in the mid-19th century was not surpassed until the 1950s. This early intensification would have reduced weed growth through the effects of fallowing and hoeing, but probably had little overall effect on the composition of arable plant communities.

Recession
At the end of the 19th century there was an end to agricultural affluence, as markets were overwhelmed by cheap grain from the rapidly expanding American corn belt. Britain's agriculture suffered as a result of our industrial success, which enabled the country to afford imported grain and to rely less on

Agricultural weeds of Leicestershire recorded by William Pitt in 1809

PITT'S NAME	PRESENT-DAY COMMON NAME (where different)	SCIENTIFIC NAME
The worst!		
Dog's Couch	Common Couch	*Elytrigia repens*
Benty Couch	Black Bent	*Agrostis gigantea*
Common Thistle	Creeping Thistle	*Cirsium arvense*
Spear Thistle		*Cirsium vulgare*
Chickweed		*Stellaria media*
Ivy-leaved Chickweed	Ivy-leaved Speedwell	*Veronica hederifolia*
Fat-hen		*Chenopodium album*
Willow-weed Redshank		*Persicaria maculosa*
Bird's Lake-weed	Common Knotgrass	*Polygonum aviculare*
Shepherd's-purse		*Capsella bursa-pastoris*
Wild Mustard	Black Mustard	*Brassica nigra*
Wild Rape	Charlock	*Sinapis arvensis*
Wild Radish		*Raphanus raphanistrum*
Corn Chamomile		*Anthemis arvensis*
Corn Marigold		*Chrysanthemum segetum*
Corn Crowfoot	Corn Buttercup	*Ranunculus arvensis*
Sow-thistles		*Sonchus* spp.
Less important		
Nettle Hemp	Common Hemp-nettle	*Galeopsis tetrahit*
White Dead-nettle		*Lamium album*
Bindweed	Field Bindweed	*Convolvulus arvensis*
Bearbind	Black-bindweed	*Fallopia convolvulus*
Shepherd's-needle		*Scandix pecten-veneris*
Corn Scabious	Field Scabious	*Knautia arvensis*
Blue Bottles	Cornflower	*Centaurea cyanus*
Colts-foot		*Tussilago farfara*
White marigold	Ox-eye Daisy	*Leucanthemum vulgare*
Groundsell	Groundsel	*Senecio vulgaris*
Scorpion-grass	Field Forget-me-not	*Myosotis arvensis*
Goose Tansy	Silverweed	*Potentilla anserina*
Corn Mint		*Mentha arvensis*
Two-seeded Tare	Hairy Tare	*Vicia hirsuta*
Four-seeded Tare	Smooth Tare	*Vicia tetraspermum*
Corn Goosegrass	Corn Cleavers	*Galium tricornutum*
Nettles		*Urtica dioica*
Poppy	Common Poppy	*Papaver rhoeas*
Cockle	Corncockle	*Agrostemma githago*

home produce. The agricultural recession lasted from the 1880s until the end of the 1930s, when self-sufficiency became a priority during the Second World War. This recession had a massive impact on the countryside, with large areas of arable land dropping out of cultivation and many farmers going out of business. This period included the start of agricultural mechanisation, the internal combustion engine being employed first in America, with early tractors, reaper-binders and, eventually, combine harvesters.

The second agricultural revolution – reforms

Amongst the measures of social reform introduced by Attlee's Labour government immediately after the Second World War, was the Agriculture Act of 1947. As well as other provisions, this guaranteed prices and markets for produce, with the aim of ensuring the more efficient use of Britain's agricultural land. The memory of the recent war concentrated people's minds on the importance of self-sufficiency in food. There were many parallels with the Corn Laws, and the Agriculture Act led to the rapid intensification of all aspects of farming practice. In contrast to the 19th century situation, however, farming subsidies in the mid-20th century were accompanied by huge

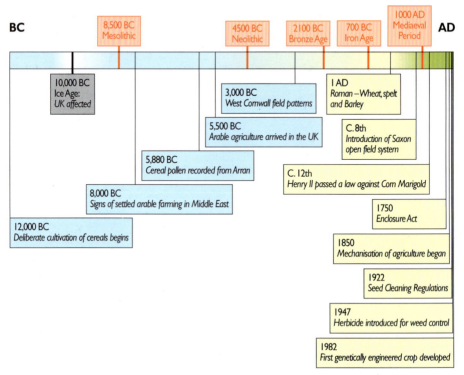

Notable events in the history of agriculture in Britain.

technological advances. The introduction of the Common Agricultural Policy (CAP), following Britain's accession to the Common Market in 1973, led to further intensification of farmland management. Minimum prices were set by guaranteeing markets for surplus production (which was put into storage), subsidies were paid to producers and exporters, and tariffs were set for imports.

Innovations

Nitrogen – no contest
In the mid-19th century, Lawes and Gilbert established several experiments on Lawes' farm in Hertfordshire (now Rothamsted Experimental Station) and convincingly demonstrated that plants needed nitrogen and other nutrients. These experiments were initially set up to provide support for Lawes' fertiliser

Nitrogen fertiliser is not directly toxic to arable plants, but has its effect by increasing competition between the crop and other 'non-target' species. Where rainfall is sufficient, as in northern Europe, competition is mainly for light beneath the densely-shading crop canopy. Experiments in Hampshire in 1988 showed how a modern winter Wheat variety fertilised at a level typical of farming practice can reduce the numbers of uncommon arable plant species. Lamb's Succory and Mousetail were not present at all where the full amount of nitrogen had been applied, whilst Small Alison, Broad-leaved Cudweed, Rough Poppy, Prickly Poppy, Corn Buttercup, Shepherd's-needle and Broad-fruited Corn-salad were all much less abundant. Field Gromwell and Thorow-wax, on the other hand, were little affected.

Species	Numbers of plants per m²		
	No nitrogen	Half-application	Full application
Small Alison	10·5	6·8	2·8
Corn Buttercup	12·1	7·3	6·2
Broad-fruited Cornsalad	11·7	6·8	2·1
Broad-leaved Cudweed	10·7	5·8	2·4
Field Gromwell	7·9	4·9	8·8
Mousetail	2·6	0	0
Prickly Poppy	5·3	1·6	1·1
Rough Poppy	1·8	1·0	0·4
Shepherd's-needle	7·1	4·4	3·1
Lamb's Succory	1·6	0	0
Thorow-wax	5·6	4·2	4·3

In other trials, the majority of arable plants have also been shown to be disadvantaged by high nitrogen applications, although many common arable plants respond positively. The latter include shade-tolerant plants, such as Common Chickweed, and a number of highly competitive species. The greatest weed problems of modern cereal farming, such as Black-grass, Cleavers and Wild Oat, can all take advantage of added nitrogen to the detriment of the crop.

Jean-Francois Millet's *The Gleaners*: one of his rural scenes for which he was renowned.
The Gleaners (1857) Jean-François Millet (1814-1875) Musée d-Orsay, Paris.

manufacturing business, and some are still running. Despite the increasing availability of artificial nitrogen fertilisers in the latter half of the 19th century, however, animal manures remained the chief source of nitrogen until the 1940s and the advent of mass mechanisation. The amounts of nitrogen applied to arable crops increased rapidly after 1945, increasing by 900% in winter Wheat and 500% in spring Barley by 1988.

In the mid-19th century, crop-breeding also expanded to produce new varieties which responded to increasing levels of fertiliser application and gave higher yields. The development of highly competitive varieties has accelerated rapidly since 1945, and the application of very high levels of nitrogen to crops has probably had a major effect on the composition of arable plant communities and the abundance of individual arable plants.

Power – horse, steam and diesel

One of the major innovations of the late 19th century was the use of steam power to drive ploughs and threshing machines. The former allowed more rapid ploughing of existing arable land and the cultivation of formerly unploughable land on steep slopes. The direct descendant of the steam plough was the tractor. As steam plough replaced draught horse, and tractor replaced steam plough, this evolution led eventually to the large-scale mechanisation of all farming operations and the hedge-removal campaign of the latter half of the 20th century.

Modern farm machinery.

Threshing and the combine harvester

The threshing machine was perhaps the first invention to have a serious effect on the composition of arable plant communities and to cause major declines in the abundance of individual species. Many arable plants bear their flowers at the same height as the ears of cereal plants and their seeds are therefore

harvested along with the crop. The pre-mechanisation processes of flailing and winnowing would have separated seeds that were much smaller than cereal grains, such as those of poppies, Charlock or mayweeds. Other seeds would have been stored along with the harvested grain and re-sown along with the seed corn. This would have helped to transport arable plants between fields and between farms. For some arable plants, storage and re-sowing was essential, particularly for those with little ability for the seed to persist in the soil, or with seed that was liable to rot or be eaten by pests.

Cleaner seeds

The threshing machine enabled the removal of many seed contaminants from the cereal grain by passing the harvested material through a grid which retained seed larger than the grain, and another which allowed passage of seed smaller than the grain. The best-known example of an arable plant with a short seed-dormancy and that relied upon regular re-sowing with a crop is Corncockle. This was once a notorious seed contaminant, particularly of Rye, but is now virtually gone from Britain and rapidly disappearing from the whole of Europe. Another former contaminant of Rye was Cornflower. The traditional weeds of Flax crops were similarly reliant upon being re-sown with the crop, and have now almost vanished from Europe. Seed-cleaning technology has improved steadily throughout the 20th century with the introduction of the combine harvester. Since the introduction of the Seeds Regulations in 1922, which specified the maximum permissible levels of contaminants, the Official Seed Testing Station has been analysing crop seed for this purpose.

The age of speed

The advent of the combine harvester and the tractor has meant that ploughing, seed-bed preparation and harvesting now take a fraction of the time that they would have taken in the age of the horse. Operations that used to take weeks, or even months, can now be accomplished in hours. The plough and harrow can now follow the combine harvester, leaving no time for late-flowering arable plants to grow in the stubbles. Crops are harvested earlier than in the past, and drilling occurs earlier too. This has had important consequences for plants with restricted germination periods and long growing seasons.

Herbicides

Chemical weed control began at the end of the 19th century when copper sulphate was first used for the control of Charlock. Sulphuric acid, ferrous sulphate and dinitro-*ortho*-cresol (DNOC) were added to the repertoire before 1940, and several forms of artificial nitrogen fertiliser were applied as sprays for their dual effect as a herbicide and a nitrogen source. These early chemicals all relied upon their corrosive effect on weed leaves. The first modern herbicide that acted internally on the target plants was MCPA (2-methyl-4-chlorophenoxyacetic acid), which was introduced in 1943. MCPA was originally recommended for the control of Corn Buttercup and Cornflower, amongst other arable plants. Herbicide use has been a very important factor in the decline and disappearance of many arable plants, although different species have different susceptibilities to different herbicides.

Many of the early compounds did not give complete control of their target species. Herbicides have, however, become more and more effective at smaller and smaller doses. At the beginning of the 21st century, complete annual eradication of weeds is now possible in many fields. Although most annual arable plants have seed that can remain dormant in the soil, the regular destruction of germinating seedlings has had a rapid effect on the populations of susceptible arable plants by depleting the seed-bank in the soil.

On the edge

The removal of hedgerows and other field boundaries to make room for larger machines has also had a profound effect on arable wildlife. The field boundary provides a home to many animals, invertebrates and plants. It also creates a zone of lower farming intensity that extends approximately 4 m inwards from

Changes in the arable flora in the 20th century

Surveys Information on the abundance and distribution of arable plants before the late 20th century is very patchy, and even recent data are incomplete and unreliable. Some measure of the declines may be demonstrated using information from national surveys of plant distribution. Surveys with the specific aim of searching for arable plants were carried out during the early 1970s and in 1986–7, and these added much valuable data to national surveys.

20th century extinction Some arable plants, including Corn Woodruff, Violet Horned-poppy and the whole community of Flax weeds became extinct early in the 20th century, and a further eight species have become extinct in recent years. Where they still occur in mainland Europe, Lamb's Succory, Downy Hemp-nettle, Small Bur-parsley and Thorow-wax are typical of communities of the highest conservation value. Darnel was once an abundant grass in Britain, and it is still present on the Aran Islands off the west coast of Ireland, where it grows in tiny fields of Rye and Potatoes and on the roofs of houses thatched with Rye straw.

Violet Horned-poppy –
Extinct in the UK

On the verge Corn Buttercup, Spreading Hedge-parsley, Small-flowered Catchfly and Broad-fruited Cornsalad are now in a very perilous situation. Some geographically restricted species have, however, maintained their core distributions.

Hanging on Despite much reduced population sizes, Field Gromwell, Rough Poppy, Prickly Poppy, Narrow-fruited Cornsalad, Night-flowering Catchfly and Dense-flowered Fumitory can still be found together in some chalky fields from Dorset to the Norfolk coast. Small-flowered Buttercup is still widespread around the coasts of south-west Britain, but rarely now in arable land. Corn Parsley occurs on both arable and non-arable sites, especially in coastal areas, and is now endangered in the rest of Europe. Some arable plants, such as Field Cow-wheat and Ground-pine, were once more frequent in arable land but now rarely occur in such habitats in Britain.

Flax Dodder –
Extinct in the UK

A very few rare arable plants seem to have increased in recent years. The only ones that have are Shepherd's-needle and Lesser Quaking-grass, Dense Silky-bent and Nit-grass. The apparent increase of these grasses is probably the result of better recording methods.

The current status of some species is uncertain. Several of these, including Rye Brome, Field Brome and Corn Chamomile, are difficult to identify and this may account for our poor knowledge of their distribution. The status of other arable plants would have been similarly poorly known had it not been for recent intensive data collation work.

Rough Poppy –
Rare but hanging on

Corn Chamomile –
Status uncertain

the field edge. Here, herbicides and fertilisers are less efficiently applied and the soil is compacted by turning machinery, leading in some cases to crop yields as little as half those of the rest of the field. In a typical modern field, most arable plants are confined to this field headland zone, with only a few species being present further into the field. As hedgerows have been removed, so have these field margin refugia, exposing the surviving flora to the full rigours of herbicide and nitrogen application. Nearly 50% of British hedgerows were removed between 1945 and 1991, aided by grants from the government and the European Union.

Non-stop cropping
The availability of chemical herbicides and artificial fertilisers has freed farmers from the constraints of crop rotations, enabling the same crop to be grown in a field for successive years. In large areas of eastern England, winter crops are grown continuously, leading to a loss of spring-germinating weeds, but resulting in the build-up of autumn-germinating species which can tolerate both the herbicides used and the high competitive pressure from the heavily fertilised crop. Black-grass, Barren Brome and Cleavers thrive under such régimes and all pose problems in modern agriculture. In some places, herbicide-resistant strains of Black-grass have become prevalent. Paradoxically, Shepherd's-needle, a rare plant in much of Britain, has undergone a resurgence in parts of East Anglia under intensive winter cropping. The arable system in this area relies upon the use of particular herbicides to which Shepherd's-needle is resistant.

Autumn-germinating and winter-germinating arable plants might have been expected to fare better with the increase in autumn-sown crops. However, many such species are those that have declined most. The reasons for this are that:

▶ autumn-sown crops are grown more intensively than spring-sown crops, with higher inputs of nitrogen and herbicides; and
▶ autumn-germinating arable plants tend to have simpler dormancy mechanisms and shorter-lived seed.

Numerous other innovations and changes have also contributed to the overall effects of intensification on farmland wildlife in general and on arable plants in particular. These include:

▶ improved field drainage;
▶ stubble burning;
▶ the direct drilling of crops; and
▶ the reduction in the variety of crops used.

The third agricultural revolution – Agri-environment schemes
By the mid-1980s, the costs of maintaining the Common Agricultural Policy (CAP) had increased to such an extent that the need for reform was acknowledged. The effects of the CAP, and the consequent intensification of

farming practice on environmental quality, were also becoming obvious throughout Europe.

Set-aside was introduced as a production control measure throughout the European Union (EU) in 1992, after a pilot scheme which ran from 1988 to 1992. This is a method of ensuring that farm incomes are maintained whilst attempting to reduce crop surpluses by withdrawing land from production. Although it has much potential for enhancing farm wildlife, specific environmental objectives were not integrated into the programme. Although there is some evidence that set-aside encouraged the intensification of production on other arable land, it did have some accidental benefits.

In 1987, the first *Environmentally Sensitive Areas* (ESAs) were designated. Within these areas, funding is provided to farmers and landowners to manage distinctive landscapes, wildlife habitats and historic features. Although the Breckland ESA included options for the conservation of the very specialised flora of disturbed land in this region, the major emphasis was on the conversion of arable land to grassland, which may have had a detrimental effect on farmland biodiversity in some areas.

Countryside Stewardship was piloted from 1988 and adopted as a government-run scheme in 1995. It applies to the whole of England and offers farmers a suite of flexible options to manage their land to enhance the landscape, wildlife and historic interests and improve access. Although much of the emphasis in arable areas has been on the conversion of arable land to grassland, there are options for the conservation of arable plants.

Arable Stewardship, aimed specifically at the management of arable land for biodiversity, was introduced in pilot areas in central East Anglia and the West Midlands in 1998. The results from the East Anglian pilot area in particular, where several rare arable plant species were known to be present, have been very encouraging, and the option was introduced into the mainstream Countryside Stewardship Scheme nationally in 2002.

The Countryside Stewardship and ESA schemes are partly funded by the EU and partly by the UK government. In England the schemes are administered by the Rural Development Service (RDS) of the Department for Environment, Farming and Rural Affairs (Defra). RDS Project Officers for these schemes work closely with landowners to ensure that the funds available are spent to best effect. In Wales the schemes are adminstered by the Countryside Council for Wales (CCW), in Scotland by the Scottish Executive and in Northern Ireland by the Department of Agriculture and Rural Development Northern Ireland (DARDNI).

The plight of arable plants, and the importance of arable field margin habitats has been recognised in the UK Biodiversity Action Plan (UKBAP). Wildlife species in need of conservation action are listed, and action plans have been drawn up to ensure their survival. In total, 66 flowering plants are listed, 12 of which, nearly 20%, are arable species. Three annual arable mosses, *Didymodon tomaculosus*, *Ephemerum stellatum* and *Weissia rostellata*, are also considered to be in need of such attention. A complete list of the UKBAP species of flowering plants is included in Appendices 3 and 4.

Mediaeval

The biology of arable plants

Arable land supports numerous crops under various management systems, but the one feature they all share is a very high degree of regular disturbance. Arable crops are grown on an annual cycle, and one of the aims of cultivation is to harvest or destroy all of the above-ground vegetation that appeared during the previous growing season. Normally, arable fields are cultivated every year and replanted with crop seed. In addition to this, herbicides are usually applied in an attempt to kill any plant other than the crop. An arable field is therefore not an easy place for a plant to survive. All wild plants that aspire to grow with arable crops must have a strategy not just to survive regular destruction, but also the long periods when conditions are unsuitable for growth.

Annuals
The majority of arable plants are annuals, which means that they complete their life-cycle from germination to death within one year. Many species can have a much shorter life-span: Shepherd's-purse and Groundsel can produce seeds as little as six weeks after germination, and can germinate at almost any time of the year given enough warmth and moisture. These are ephemeral plants, which can exploit a much wider range of habitats than just arable land. Most of the wild plants more strictly confined to arable land have life-cycles closely synchronised with the timing of traditional arable farming. The characteristics of these two groups are summarised below, although relatively few species in each group fulfil them all.

Perennials
Biennials and other short-lived perennials can sometimes flourish in areas that are occasionally, but not annually, disturbed by cultivation.

Such species include Perennial Sow-thistle, Creeping Thistle and Common Couch. Other plants like Common Nettle, Colt's-foot and even Bramble can become established where cultivations are shallow. The only uncommon

	Species able to grow in many habitats	Species confined to arable land
Amount of seed produced	Lots	Little
Seed size	Small	Large
Seed mobility	Mobile	Immobile
Seed bank	Short-lived	Long-lived
Germination time	Throughout the year	Restricted
Habitat	All disturbed land, often transient sites	Arable land, rarely other regularly disturbed land

perennial arable plant is Tuberous Pea, which may have been a 19th century introduction, and is confined to a small area of Essex. Perennials can often pose problems even in high-input arable systems, as they are very difficult to eradicate with herbicides. In areas managed for conservation, they can be problematic, as they rapidly form dense stands in areas without crop cover. These perennials have solved the problem of re-establishment after ploughing in two ways. Some have creeping rhizomes that can either grow beneath plough-depth, or which give rise to new plants from each fragment when broken up by the plough. Others produce small tubers, which remain in the soil after ploughing.

Perennial weeds:
Species with underground rhizomes:
Black Bent
Bramble
Colt's-foot
Common Couch
Creeping Thistle
Perennial Sow-thistle
Common Nettle
Rosebay Willowherb
Field Horsetail

Tuber-forming species:
Field Garlic
Onion Couch
Tuberous Pea

Seed production

All annual plants rely upon regular seed production to maintain their populations. Most annuals can form a persistent bank of seed in the soil. The number of seeds produced varies greatly between species and between plants of the same species. This variation depends upon the availability of nutrients and water in the soil and the amount of light available. This plasticity is typical of annual plants. A plant of Common Poppy may produce only a single seed-pod under conditions of extreme competition from the crop, whereas a nearby plant growing in a gap where the crop has failed may produce more than 100 seed-pods. Common Poppy produces particularly large numbers of seed, on average around 16,500 per plant. The now extinct Thorow-wax, on the other hand, produced only 170 seeds on average per plant. The amount of seed that is returned to the seed-bank each year is important in determining the survival of a population.

Seed-banks and dormancy

In a cultivated field, seeds do not remain on the soil surface for long after being shed from the parent plant and they are usually incorporated into the soil by ploughing or other cultivations to a maximum depth of approximately 20 cm. This can prevent immediate germination, and dormant seeds in the soil are referred to as the seed-bank. Long-lived seed is a characteristic of most arable plants, and gives obvious advantages to plants living in a habitat which may be unsuitable for long periods. Germination is stimulated by exposure to light, and in a regularly cultivated field, a fresh batch of buried seed will be exhumed every year. Weed control by bare-fallowing takes advantage of this by attempting to stimulate germination by repeated cultivations during which the already-germinated seedlings are destroyed.

The longevity of dormant seed is primarily an inherent property of each species. Some are very long-lived. Poppies, for example, are thought to be able to survive for hundreds of years. At the other extreme, the seed of Corncockle, and other species which rely upon being harvested along with the crop and being re-sown with the next season's crop, survive for just a few months. Seed longevity can be reduced by predation by birds, small mammals or soil organisms. Seeds can also be lost to fungal decomposition. Longevity can be increased, however, by deep-burial in anaerobic conditions, and species such as Corncockle and Cornflower sometimes appear after the ground has been disturbed for foundations, pipelines or new roads in areas where they have not been seen for many years.

A question of timing

Annual arable plants are suited to growing in conditions of regular, predictable disturbance, provided that their germination period occurs after cultivation and sowing of the crop. The majority of species that are confined to arable habitats have very restricted germination periods, either predominantly in the late autumn and early winter or in the spring. This germination time period is partly a result of a simple response to external conditions; a seed will not germinate if it is too dry or too cold. It is also a result of the internal programming of the seed interacting with external conditions, chiefly temperature. The combination of these factors means that a seed near the soil surface goes through a cycle of dormancy in which it is dormant at certain periods of the year when conditions are unsuitable for germination and can germinate during those periods when conditions are suitable.

The consequence of these restricted germination periods is that the plants germinating in a field after cultivation on one date may be very different from those germinating after cultivation on another date. Quite minor variations in the timing of cultivation and crop drilling can have major effects on the plant community. The earlier drilling of cereals in autumn and winter has had a dramatic effect on arable plant communities.

Life-cycles

The length of time between germination and seed production also varies between species. As mentioned previously, some annual plants can have a life-cycle of as little as five weeks. A longer life-cycle is more normal, however, and a very few species have life-cycles that are longer than those of the crops with which they grow.

As an example, in a winter Wheat crop drilled at the beginning of October, the bulk of the arable plants germinate before the beginning of January. Provided that the winter has not been too hard, some will start to flower by mid-May. The great majority of plants will be in flower by the middle of June and will have set seed by the end of July, before harvest. The length of the flowering period will depend to some extent upon the weather, and will be extended in a cool, wet summer.

Main germination periods of some annual arable plants

	Jan	Feb	Mar	Apr	May	Jun	Jul	Aug	Sep	Oct	Nov	Dec
Corn Buttercup										▓	▓	
White Campion			▓	▓				▓	▓			
Night-flowering Catchfly			▓	▓								
Broad-leaved Cudweed			▓						▓	▓		
Field Forget-me-not				▓	▓				▓	▓	▓	▓
Corn Gromwell				▓					▓	▓	▓	
Corn Marigold				▓	▓							
Mousetail	▓									▓	▓	
Babington's Poppy				▓				▓				
Common Poppy				▓	▓							
Long-headed Poppy				▓					▓	▓	▓	▓
Prickly Poppy				▓					▓	▓	▓	▓
Rough Poppy				▓					▓	▓	▓	▓
Shepherd's-needle				▓						▓		
Lamb's Succory				▓					▓			
Weasel's-snout					▓							

Nitrogen inputs and competition with the crop

Modern crop varieties have been bred to be highly responsive to nitrogen fertiliser. They rapidly form a dense canopy of leaves after germination that shades out the seedlings of other slower-growing arable plants. The majority of the arable plants that have decreased in recent years are relatively slow-growing, and most have a relatively low stature, rendering them vulnerable to the effects of competition with the crop. The taller-growing plants such as Cornflower, Common Poppy and mayweeds eventually reach the crop canopy, but even these cannot compete with the selectively-bred crop plants to take up the fertiliser. Where such species appear to flourish, it is usually because the crop has failed for some reason.

Herbicides and herbicide resistance

Modern herbicides operate on specific biochemical reactions within plants. The earliest-developed compounds such as MCPA work by imitating the actions of natural plant hormones. Many of the most recently developed compounds operate on single steps in photosynthesis. Arable plants vary greatly in their susceptibility to herbicides. This is partly due to internal biochemical differences, such as those that separate grasses from broad-leaved plants. Physical factors can also be important in preventing the uptake of herbicides by plants. In general, the broader a leaf is, the more easily the herbicide is absorbed. Features that can hinder absorption include hairy leaves and the presence of a waxy leaf cuticle. Corn Marigold, for example, has very waxy leaves, and most water-based herbicides simply run off. The decline of this plant can be partly explained by the development of efficient wetting agents during the 1970s which helped in the retention of herbicide spray on the leaves.

The prolonged use of a single herbicide or group of herbicides can lead to the development of herbicide-resistant strains of some plants. Herbicide-resistant Black-grass is a particular problem in winter cereals in southern England.

Biotechnology has sought to introduce genes for herbicide-resistance into crop plants. The herbicides that are being investigated are very broad-spectrum compounds including Glyphosate. The widespread use of such chemicals in growing crops could pose a very serious threat to the arable flora, and could lead to the evolution of strains of some problem species that are more herbicide-resistant.

Uncommon **early-flowering plants** include Mousetail, Corn Buttercup and Shepherd's-needle.

Some plants **flower later**, and mayweeds, in particular, can be a feature of a winter Wheat crop in August. Spreading Hedge-parsley and Broad-leaved Spurge are rare arable plants that flower late.

Some arable plants are rarely able to set seed in the standing crop, and depend upon the stubble being left after harvest. Corn Parsley germinates in late September and early October, begins to flower in late July, and is usually unable to produce much seed before harvest. However, the lower branches often escape the combine harvester and produce seed in the stubble in August and September.

Other plants which flower best in **stubbles** are Red Hemp-nettle, Night-flowering Catchfly and the fluellens, all of which are more characteristic of spring-sown crops. This reliance upon stubbles is partly due to the natural late-flowering of these plants, but is also a result of the sudden freedom from the constraints of the crop after harvest. Stubble is also good for mosses and liverworts.

The good, the bad and the ugly

The decline of most arable plants, the continued persistence of others and the rise to prominence of a few can be explained by their biology. The ability to form persistent seed-banks is an obvious advantage to an arable plant, and can provide a buffer against the effects of changes in arable farming practice. Many of those species that have declined so dramatically have relatively short-lived seed-banks, conferring little resilience on populations of these species. Corncockle is an extreme example, and other species with short-lived seed, such as Cornflower, Corn Buttercup and Shepherd's-needle, declined rapidly during the 1950s and 1960s as herbicides became more effective and crops more competitive. Susceptibility to herbicides has been the feature that has rendered species most liable to rapid decline, although their inability to compete with a highly fertilised crop has also been very important.

Arable plants such as Common Poppy, knotgrasses and Charlock that have persisted, even if in much reduced quantities, have been those which,

Some reasons for the decline of arable plants of conservation concern

	Herbicides	Nitrogen & competition	Changes in timing	Drainage	Crop seed cleaning	Short-lived seed
Interrupted Brome		◆	◆		?	◆◆
Corn Buttercup	◆◆	◆◆	◆			◆
Small-flowered Catchfly	◆◆	◆◆	◆			
Corn Cleavers	?	◆◆	?			◆◆
Broad-fruited Cornsalad	◆◆	◆◆				
Narrow-fruited Cornsalad	◆◆	◆◆				
Cornflower	◆◆	◆			◆	◆
Broad-leaved Cudweed	◆◆	◆◆				
Red-tipped Cudweed	◆◆	◆◆				
Grass-poly	◆◆	◆		◆◆		
Field Gromwell	◆◆		◆			?
Spreading Hedge-parsley	◆◆	◆◆	◆◆			
Red Hemp-nettle	◆◆	◆◆	◆◆			
Corn Parsley	◆◆	◆	◆◆			
Pheasant's-eye	◆◆	◆◆				
Martin's Ramping-fumitory	◆	◆				
Purple Ramping-fumitory	◆◆	◆				
Western Ramping-fumitory	◆◆	◆				
Shepherd's-needle	◆◆	◆				◆
Fingered Speedwell	◆	◆◆				
Broad-leaved Spurge	◆◆	◆				

although highly sensitive to herbicides, have long-lived seed and a buffering seed-bank, replenished by occasional chance events that allow seed production. Those species which have become the modern weeds, such as Barren Brome, Cleavers and Black-grass, are opportunists, pre-adapted to the conditions of contemporary cereal farming. Although they have little seed-dormancy they are resistant to many herbicides and are highly responsive to nitrogen. They germinate early in the autumn and produce seed early in the summer, synchronising well with modern cropping cycles.

An understanding of some of these features of the basic biology of arable plants can help us understand the factors that have led to their decline or in some cases to their increase. More importantly, it may tell us how we can approach their conservation.

Farming and wildlife

John Davis

How to use this guide

This guide is designed to help identify arable plants in the field, as well as provide information which will help farmers and other landowners manage for these plants. If you are still unsure about a plant's identification, contact someone (see pages 305–308); it is against the law to uproot plants unless on your own land.

Scope of Book

This book covers the rare plants found on arable land including arable and arable-edge species listed in the *British Red Data Book (Vascular Plants)*, *Scarce Plants in Britain* (see Appendix 1) and the *UK Biodiversity Action Plan (UKBAP)*. A few additional species are included which are in serious decline.

Order

The plants are listed in alphabetical order with similar species of *e.g.* buttercup, cornsalad, fumitory or poppy grouped together. NB: Family order is followed in other identification guides. The flowering plants are outlined before the grasses and mosses. Names of plants and grasses follow Stace's *New Flora of the British Isles* 2nd Edition (1997). English names may differ between regions and countries, but the scientific name stays the same.

Species accounts: arable plants

Each species is described and illustrated by a *photograph* taken in the wild. *Line drawings* of seeds or leaf shapes are provided to help with identification of closely related plants. A *distribution map* indicates where the plant has been known to occur in Britain recently whereas the text on the plant's distribution outlines the current state of knowledge. The *flowering time* is suggested, but this will vary and needs interpreting flexibly. A cold spring will delay flowering, while plants in Cornwall tend to flower in the UK earlier than those in eastern England or in the north.

Other plants on arable land

The species accounts include an outline of differences to look for between the species being described and similar looking ones, some of which are not rare. Both the notes under **Similar species** and the **Identification Keys** to flowering plants (pages 228–243) and grasses (pages 264–265) aim to clarify the features to look for.

Some of the other species which occur more commonly on arable land are listed in Appendix 6.

There are also many non-arable plants which occur in arable habitats. These colonise from hedgerows, nearby grassland or woodland habitats. Detailed descriptions of both these species groups are beyond the scope of this book and the use of a field guide to British wild plants is recommended to help identification of these. See the bibliography on page 301.

BAP species

If you find some rare species be sure to let the local BSBI recorder (page 305) know so they can ensure that the plant is recorded for your county.

A guide to the species accounts

STATUS

The status of the species in Britain indicates whether it is **'nationally rare'** (present in 15 or fewer 10 km national grid squares) or **'nationally scarce'** (present in 100 or fewer 10 km national grid squares). Its presence on the UKBAP lists of **'Priority Species'** or **'Species of Conservation Concern'** is also noted.

English name

Scientific name

IDENTIFICATION:

A brief description of the fully-grown plant, covering the general form, shape and size of the plant, leaf structure and appearance, and the salient characteristics of the flower, fruit and seed important to its identification.

Similar species: The salient differences between confusing species. Further details of these species can be found in the **Identification Keys** on pages 228–243 (plants) and 264–265 (grasses).

Associated uncommon species: Other plants that the species regularly occurs with.

HABITAT:

A concise description of the species' preferred habitats.

SOIL TYPE:

The soil types on which the species is found.

MANAGEMENT REQUIREMENTS:

Notes on the management techniques that the species needs in order to prosper where these are specific to the individual species and over and above the general management principles of arable plants.

DISTRIBUTION MAP

Map showing the *historical* distribution of the species, indicating where plants were recorded in Britain and Northern Ireland between 1971 and 1987.

DISTRIBUTION:

A description of the *current known* distribution, and any relevant historical information.

LIFE CYCLE:

Notes on flowering period, seed biology and germination strategy with chart indicating flowering (green) and germination (brown) periods.

J F M A M J J A S O N D

PLATES AND ILLUSTRATIONS

Each plate has a photograph of the species and, wherever possible, depicts its flower, leaf and seed. In addition, most species are accompanied by a detailed illustration, drawn to scale, of any important features that aid identification.

REASONS FOR DECLINE:

A paragraph detailing those factors thought to be responsible for the decline of the species.

Deciding which plant it is

The description in the text is designed to help work out which plant you have found. Although plants have individual features which can help differentiate between species, these differences can be small and variable. As a general rule, useful things to note include:

▶ the stem – is it smooth or hairy and round or square?

▶ the leaves – what shape are they, what form do they take, where are they on the plant?

▶ the flowers – what is their colour, size and shape and how many petals do they have?

▶ the fruit and seed – what shape and size are they?

Other factors. such as soil type, fertiliser/herbicide use will affect the variety of species that occur.

Identification aids

Some species, such as the cornsalads and fumitories have very subtle differences. A magnifying glass can be extremely useful in helping to distinguish betwen them.

To help in measurement in the field, the inside cover of this book has a ruler scale along its long edge.

Plant terminology

The terms used to describe parts of plants can be difficult to interpret without experience. Wherever possible, this guide has tried to avoid using these terms. Where they are used there is an explanation of their meaning in the following *Glossary*, along with diagrams indicating their location, where relevant, on both a flowering plant and a grass.

Plant Structure

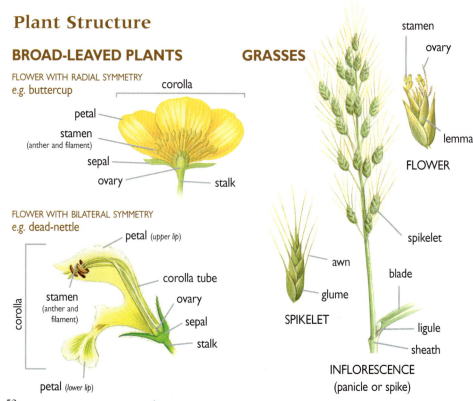

BROAD-LEAVED PLANTS

FLOWER WITH RADIAL SYMMETRY
e.g. buttercup

corolla
petal
stamen
(anther and filament)
sepal
ovary
stalk

FLOWER WITH BILATERAL SYMMETRY
e.g. dead-nettle

petal (*upper lip*)
corolla tube
corolla
stamen
(anther and filament)
ovary
sepal
stalk
petal (*lower lip*)

GRASSES

stamen
ovary
lemma
FLOWER

awn
glume
SPIKELET

spikelet
blade
ligule
sheath

INFLORESCENCE
(panicle or spike)

Glossary

anther	The part of the flower, at the end of a stalk (filament), that produces the pollen.
appressed	Flattened against but not joined to.
auricle	Small projections at the base of a leaf (especially GRASSES).
awn	A stiff bristle projecting from the **glume** or **lemma** in grasses.
axil	The junction of leaf and stem.
basal rosette	A circle of leaves that forms close to the ground.
bract	A small leaf-like organ from which the flower arises and which may or not be also a **bracteole**.
bracteole	A **bract** directly below the flower (especially in the carrot family APIACEAE).
calyx	the ring of, usually green, petal-like **sepals** just below the actual **petals**.
corolla	the **petals** as a whole unit, especially when they are joined, often as a tube-like structure.
floret	A small flower that is part of a compound flower-head.
fruit	The organ that contains the seeds.
glumes	The two **bracts** at the base of a **spikelet** in grasses.
inflorescence	The part of a plant that bears the flowers.
lanceolate	Spear-shaped, narrowly oval and pointed (LEAVES).
leaflet	One part of a compound leaf (LEAVES).
lemma	The lower of two **bracts** enclosing the flower in grasses.
ligule	The growth, usually membraneous, at the inner junction of the leaf-sheath and blade in grasses.
lobe	Part of a divided leaf, although not divided into separate **leaflets**.
midrib	The large, strengthened vein from the stem to the tip of a leaf.
ochrea	A sheath at the base of the leaf forming a tube around the stem (plural ochreae) (especially KNOTRASSES, *POLYGONUM* SPP.).
ovary	The part of the plant that develops into the seed.
panicle	The branched flower-head of a grass.
pappus	An arrangement of fine hairs.
petal	One part of the flower or **corolla**.
pinnate	Divided into separate **leaflets** (LEAVES).
raceme	A long **inflorescence** with flowers on either side.
ray	One stalk of an **umbel**.
ray-floret	An individual **floret** that looks like a **petal**.
sepal	One part of the **calyx**.
spike	A long **inflorescence** with flowers arising from a central stalk.
spikelet	The base unit of the **inflorescence** containing one or more flowers in grasses.
spur	A tubular projection from a **petal** or **sepal**.
stamen	The male organ of a flower comprising an **anther** on a stalk (or filament).
stipule	A scale- or leaf-like organ at the base of a leaf stalk.
umbel	A flower-head shaped like a flattened umbrella.
whorl	Three or more flowers or leaves arising from the same part of the stem.

Small Alison

Alyssum alyssoides

IDENTIFICATION:

Small Alison is an erect plant branched from the base, that grows up to 30 cm tall, although often much smaller. The leaves are up to 2 cm long, spear-shaped, narrower at the base than at the tip and covered with minute star-shaped hairs, giving the plant a grey appearance. The flowers are borne in a terminal spike. They are small, up to 3 mm in diameter, and have four yellow petals which fade to white as the fruit is produced. The fruit is spherical and about 3 mm in diameter, containing four orange-brown seeds.

Similar species: No other plant in the cabbage family (Brassicaceae) has star-shaped hairs and tiny yellow flowers. The shape of the fruit is diagnostic.

Associated uncommon species: At its East Anglian sites it has been recorded with Perennial Knawel, Sand Catchfly, Small Medick and Toothed Medick.

HABITAT:

Non-competitive arable crops, tracksides and sand pits.

SOIL TYPE:

In Britain, Small Alison has normally been a species of sandy soils, but in the rest of Europe it is often found on thin, stony limestone soils.

MANAGEMENT REQUIREMENTS:

Occasional cultivation of field edges and disturbance of tracks where it occurs in such habitats. Occasional fallowing of arable fields.

DISTRIBUTION:

In the past this species was found at many sites across the whole of the south of England. In recent years it has been found only in East Anglia, with records only from one site in Breckland and one in the Suffolk Sandlings.

LIFE CYCLE:

Flowers from May to June.
Seed is relatively short-lived.
Germination is mainly in the autumn.

J F M A M J J A S O N D

REASONS FOR DECLINE:

This may have been a species that relied at least partially on introduction with small-seeded crops. It has been affected by intensive cultivation of field edge habitats, increases in the use of nitrogen and the development of more competitive crop varieties.

Small Alison: fruit ×5

Four-leaved Allseed

Polycarpon tetraphyllum

IDENTIFICATION:

Four-leaved Allseed is a low-growing, often prostrate plant, with a much-branched stem, up to 15 cm tall. The leaves are oval, broader near the tip than at the base, and with rounded ends. The lower pairs of leaves appear to be arranged in whorls. The flowers are small, 2–3 mm across. The 1 mm long white petals, are shorter than the pointed sepals. The petals fall early, and the flowers sometimes never open at all. Flowers are borne in clusters at the ends of stems.

Similar species: Other species in the pink family (Caryophyllaceae) with small, white flowers, such as Chickweed, mouse-ears and spurreys. Four-leaved Allseed may be distinguished from all of these by the almost whorled arrangement of the leaves.

Associated uncommon species: Other Isles of Scilly bulb field species including Prickly-fruited and Small-flowered Buttercups, Small-flowered Catchfly, Corn Marigold, Field Woundwort, Corn Spurrey, Common Cornsalad, Field Madder and Musk Stork's-bill.

HABITAT:

On the Isles of Scilly, Four-leaved Allseed is widespread in bulb fields and other horticultural habitats. It can also be found in gardens, on sand dunes and at the base of walls. In mainland Britain it is restricted to a few exposed, sunny, south-facing banks, and on compacted shingle at its Dorset locality. It is possible that these warm, sunny sites were the original habitat of this plant.

SOIL TYPE:

Sandy soils, sometimes very stony soils.

MANAGEMENT REQUIREMENTS:

Continuation of traditional bulb-growing practice with a minimum of herbicide use, especially in the field margins.

Four-leaved Allseed: fruit ×7

DISTRIBUTION:

As an arable species, Four-leaved Allseed is restricted to the Isles of Scilly, where it is widespread and locally frequent. It is also found at a few localities along the south coast from Cornwall to Dorset.

LIFE CYCLE:

Flowers from June to August.
Seed longevity is unknown.
Germination is said to be largely in the spring, but the habitat and distribution suggest that it is more likely to be in the autumn.

J F M A M J J A S O N D

REASONS FOR DECLINE:

It is still a frequent species on the Isles of Scilly. It could be threatened by the excessive use of herbicides in the bulb fields or more seriously by the abandonment of traditional horticulture.

Black-bindweed

Fallopia convolvulus

IDENTIFICATION:
A slender, scrambling plant sometimes much-branched and often twining clockwise around the stems of crops and other weed plants. Where there is nothing to climb, the stems are prostrate. Stems grow up to 120 cm in length. The leaves are 2–6 cm long, heart-shaped, pointed at the apex and with a short stalk. The flowers are in loose clusters of 2–6 in the axils of the leaves. They are small and inconspicuous with greenish-white 'petals'. Each flower produces a single large seed which projects from the shrivelled flower. The seed is approximately 5mm long, triangular and black.

Similar species: The only similar species are Field Bindweed and Hedge Bindweed. Both these species have large, conspicuous flowers and twine in an anti-clockwise direction. When not flowering, Field Bindweed can be distinguished by its grey-green round-tipped leaves and Hedge Bindweed by its much larger leaves, up to 15 cm long.

Associated uncommon species: Black-bindweed can often be found in communities that include uncommon species on limestones and sands.

HABITAT:
Arable field margins and other disturbed sites.

SOIL TYPE:
Largely on well-drained soils, largely on sands or limestones.

DISTRIBUTION:
Throughout Britain. It has declined in recent years particularly in the north.

LIFE CYCLE:
Flowers from July to October.
Seed is thought to be very long-lived.
Germination is almost entirely in April and May.

J F M A M J J A S O N D

REASONS FOR DECLINE:
This was formerly a major contaminant of cereal seed, and improved seed-cleaning has probably been a factor in its long-term decline. The recent decline is due to a combination of the high levels of nitrogen applied to competitive modern crops, use of broad-spectrum herbicides and increases in the proportion of the cropped area sown with winter cereals.

Black-bindweed: seed ×6

Small Bur-parsley

Caucalis platycarpos

IDENTIFICATION:

Small Bur-parsley is a small erect annual of the carrot family (Apiaceae) growing up to 30 cm tall. The leaves are 3-pinnately divided into narrow segments. The entire plant is scattered with bristles. The white-petalled (sometimes tinged pink) flowers are clustered in umbels without leafy bracts at the base. The umbels each have 2–5 branches. The large seeds are formed in pairs, splitting apart as they are shed from the plant. They measure up to 13 mm in length, and have long, curved spines.

Similar species: Several other members of the carrot family are similar. On arable land, particularly when in flower, it can be confused with Shepherd's-needle, Spreading Hedge-parsley, Wild Carrot and Fool's Parsley, but it can be distinguished from all of these by its very distinctive seeds.

Associated uncommon species: In central Europe it is a characteristic species of the richest communities of calcareous soils.

HABITAT:
Arable field edges.

SOIL TYPE:
Calcareous clay loams.

MANAGEMENT REQUIREMENTS:
Autumn cultivation.

E?

DISTRIBUTION:
Formerly present in scattered localities on the chalk of southern and eastern England, most particularly on the East Anglian chalk. There have been no records since 1962.

LIFE CYCLE:
Flowers from late May to early July. Seeds live in the soil for at least three years.
Germination is in the autumn.

J F M A M J J A S O N D

REASONS FOR DECLINE:
It has been seriously affected by the development of competitive crop varieties and increases in nitrogen applications. It is susceptible to herbicides. The early decline may have been due to improvements in seed-cleaning technology at the end of the 19th century.

Small Bur-parsley: fruit ×2

Corn Buttercup

Ranunculus arvensis

IDENTIFICATION:

Corn Buttercup plants can grow to 50 cm in height, and are often much-branched. The leaves are stalked and divided to the base into 3–5 lobes. The solitary flowers of Corn Buttercup resemble those of other buttercups (Ranunculaceae), although they are smaller, up to 12 mm in diameter, and paler lemon-yellow in colour. The seeds are conspicuous and very characteristic, up to 8 mm in length, ovate, and covered in spines up to 2 mm in length. Each flower has up to eight seeds.

Similar species: Creeping, Small-flowered and Hairy Buttercups sometimes occur on arable land, but none have spiny seeds. The seeds of Prickly-fruited Buttercup have a spine-free edge.

Associated uncommon species: It occurs at some sites with Shepherd's-needle, Spreading Hedge-parsley, Broad-leaved Spurge and Broad-fruited Cornsalad.

HABITAT:

Almost entirely on arable land, but Corn Buttercup sometimes occurs as isolated plants on road verges where these pass through former arable land.

SOIL TYPE:

Corn Buttercup is most frequently found on heavy clay soils.

MANAGEMENT REQUIREMENTS:

Autumn cultivation.

DISTRIBUTION:

Formerly widespread and locally abundant throughout the south and east of England, Corn Buttercup has declined rapidly since the 1940s. Viable populations on arable land are rare. There is a stronghold in the south-west Midlands, with sites scattered from Devon to Suffolk.

LIFE CYCLE:

Flowers from early May to mid-June. Seed is short-lived on regularly cultivated land, deeply-buried seed may survive longer.
Germination is in autumn and winter.

J F M A M J J A S O N D

REASONS FOR DECLINE:

Corn Buttercup is susceptible to most broad-spectrum herbicides. Increasing nitrogen application and the development of nitrogen-demanding crop varieties are likely to have played a role in its decrease.

Corn Buttercup: seed ×4

Hairy Buttercup
Ranunculus sardous

IDENTIFICATION:
An erect, branched plant, typically growing no taller than 30 cm, usually with several hairy flowering stems arising from a basal rosette. Although normally an annual, it can be perennial in grassland. The leaves are hairy and toothed. Rosette leaves are each divided into three main lobes, with a stalked central leaflet. Upper leaves are less deeply lobed and have shorter stalks. The flowers, 12–25 mm in diameter, are at the ends of the branched stems. The sepals are bent back when the pale yellow flowers open. The round seeds are 2–3 mm in diameter and lack spines, although a few warts may be present.

Similar species: Other buttercups, especially Bulbous, Creeping, Meadow and Corn Buttercups. The flowers of Bulbous, Creeping and Meadow Buttercups are a more vivid yellow. Bulbous Buttercup also has its sepals bent back but differs in having a corm-like tuberous stem at its base which is diagnostic. Creeping and Meadow Buttercups do not have bent sepals. Corn Buttercup lacks a basal rosette of leaves, has small lemon-yellow flowers and large seeds with hooked spines.

Associated uncommon species: It can be found with species-rich arable plant communities at some sites.

HABITAT:
Damp arable field margins, tracks, disturbed grassland, coastal grazing marsh.

SOIL TYPE:
Sandy clays which are seasonally wet.

DISTRIBUTION:
Hairy Buttercup is widespread around the coasts of Britain, particularly in the south and east of England.

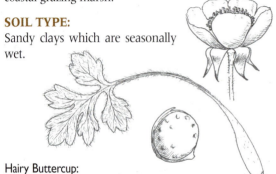

Hairy Buttercup:
leaf (*left*) ×0·5, seed (*below*) ×6, flower (*above*) ×1

LIFE CYCLE:
Flowers from early June to October. Seed biology is little-known. Germination is mainly in the autumn.

J F M A M J J A S O N D

REASONS FOR DECLINE:
Hairy Buttercup is probably susceptible to many herbicides.

64

Prickly-fruited Buttercup
Ranunculus muricatus

IDENTIFICATION:
A typical buttercup, growing up to 50 cm tall, but usually shorter. The lower leaves are long-stalked and divided into 3–5 lobes, the stalks of the higher leaves becoming gradually shorter and lobes becoming fewer. The flowers are up to 15 mm in diameter and have yellow petals slightly longer than the sepals which are bent back towards the stem. The seeds are very distinctive, 7–8 mm long with a hooked beak at one end. They are covered with long hard spines, but have a spine-free edge.

Similar species: The prickly seeds distinguish this species from all buttercups other than Corn Buttercup, which differs in having deeply-divided leaves and a spiny-edged seed.

Associated uncommon species: Other Isles of Scilly bulb field species including Small-flowered Buttercup, Four-leaved Allseed, Corn Marigold, Field Woundwort, Small-flowered Catchfly, Corn Spurrey, Common Cornsalad, Field Madder and Musk Stork's-bill.

HABITAT:
In bulb fields.

SOIL TYPE:
Sandy loams.

MANAGEMENT REQUIREMENTS:
Continuation of traditional bulb-growing practice with a minimum of herbicide use, especially in the field margins.

DISTRIBUTION:
In Britain, found only on the Isles of Scilly.

LIFE CYCLE:
Flowers from June to July.
Seed is relatively short-lived.
Germination is unknown, likely to be mainly in the autumn.

J F M A M J J A S O N D

REASONS FOR DECLINE:
This species may be threatened by the use of herbicides in the bulb fields or, more seriously, by the abandonment of traditional horticulture.

Prickly-fruited Buttercup: seed × 4

Small-flowered Buttercup
Ranunculus parviflorus

IDENTIFICATION:
Small-flowered Buttercup is normally a spreading species with branching stems up to 40 cm long. The leaves are yellowish-green in colour, roundish in shape, shallowly divided into toothed lobes and with hairy stems. The flowers are small, with pale yellow petals up to 3 mm long, and hairy sepals bent back towards the stem. The fruits are distinctive, about 2·5–3·0 mm long with a short beak and covered in short hooks.

Similar species: Other arable buttercups have larger, more vivid yellow flowers and distinct differences in the shape of their seeds.

Associated uncommon species: On arable land it can be associated with species including Weasel's-snout, Corn Parsley and uncommon fumitories.

HABITAT:
Arable field margins near the coast, horticultural land, maritime grassland and tracksides.

SOIL TYPE:
Calcareous and non-calcareous clay loams, sandy loams.

MANAGEMENT REQUIREMENTS:
Where it occurs on the Isles of Scilly, continuation of traditional bulb-growing practice with a minimum of herbicide use, especially in the field margins.

DISTRIBUTION:
This species is most frequent in the south-west of England and Wales, particularly near the coast, although there are centres of distribution in the West Midlands, and scattered sites in the east of England as far north as Yorkshire.

LIFE CYCLE:
Flowers from May to July.
Seed is thought to be long-lived.
Germination is mainly in the autumn.

J F M A M J J A S O N D

REASONS FOR DECLINE:
Probably the use of more competitive crops and the increased levels of nitrogen application.

Small-flowered Buttercup:
seed head (*left*) ×4, seed (*right*) ×6

Wild Candytuft

Iberis amara

IDENTIFICATION:

An erect plant with a much-branched stem, growing up to 40 cm tall. The lower leaves are spoon-shaped, while the upper leaves are lanceolate, broader near the tip than the base, with 2–4 blunt teeth. The flowers are up to 8 mm across and have four white petals of unequal size. The flowers are in open clusters at the ends of branches, which elongate as the fruit ripen. The fruits are circular, approximately 4–5 mm in diameter, with projecting wings that are triangular lobe-shaped.

Similar species: Wild Candytuft could be confused with other white-flowered species of the cabbage family (Brassicaceae), but the shape of its fruit is characteristic.

HABITAT:

Rabbit-disturbed chalk grassland, quarries and occasionally cultivated arable field margins.

SOIL TYPE:

Very thin soils over chalk.

DISTRIBUTION:

It is restricted mainly to the Chilterns from Reading to Cambridgeshire, with outlying sites in Wiltshire and Surrey.

Wild Candytuft:
flower (*left*) ×3, leaf (*centre*) ×3, seed (*right*) ×3

LIFE CYCLE:

Flowers from May to August.
Seed is very long-lived.
Germination is mainly in the autumn.

J F M A M J J A S O N D

REASONS FOR DECLINE:

More intensive cultivation of field-edge habitats. There has been little decline in non-arable habitats.

Smooth Cat's-ear
Hypochaeris glabra

IDENTIFICATION:

A small, dandelion-like plant with a rosette of leaves and several erect, flowering stems that are sometimes branched. The narrow, spear-shaped leaves, with occasional teeth, are broader near the tip than at the base. They are hairless, normally 1–8 cm long, but sometimes as long as 20 cm. The stems are usually 5–15 cm tall, but can reach 40 cm, and have scale-like bracts rather than leaves. The flower-head is 0·5–1·0 cm in diameter and is made up of a cluster of tiny flowers which only open in bright morning sunlight and which close by noon. Each flower has a single conspicuous yellow petal about twice as long as broad. The reddish-brown seeds are 4–5 mm long. The seeds formed at the centre of the flower have a long beak; the outer seeds do not have a beak.

Similar species: The presence of scales on the flower stems separates Smooth Cat's-ear from all other yellow-flowered species of the dandelion family (Asteraceae) with basal leaf rosettes except Common Cat's-ear, which differs in having hairy leaves and long, narrow petals.

Associated uncommon species: Smooth Cat's-ear can occur with Red-tipped Cudweed and, in Breckland, with Breckland rarities.

HABITAT:
Disturbed sandy grasslands and heaths, arable fields with very sparse crops, abandoned arable land.

SOIL TYPE:
Non-calcareous sands.

MANAGEMENT REQUIREMENTS:
Continued disturbance of poor, sandy grasslands, maintaining cultivation of sandy arable land.

Smooth Cat's-ear: leaf (*left*) ×0·5, outer seed (*centre*) ×2, inner seed (*right*) ×2

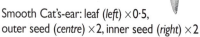

DISTRIBUTION:
Smooth Cat's-ear is scattered throughout Britain from the south coast to northern Scotland. However, it is concentrated in small areas, for example the East Anglian Breckland and Sandlings, the Surrey heathlands and the Welsh Borders.

LIFE CYCLE:
Flowers from June to October. Seed is relatively short-lived. Germination is mainly in the autumn.

J F M A M J J A S O N D

REASONS FOR DECLINE:
Increases in the levels of nitrogen applied to very nutrient-poor soils, and the use of more competitive crop varieties. Conversion of poor arable land on sandy soils to pasture, and improvement of poor pasture land.

Night-flowering Catchfly

Silene noctiflora

IDENTIFICATION:

Night-flowering Catchfly grows to about 50 cm tall with an occasionally-branched stem. The whole plant is hairy, with sticky hairs (which often catch flies!) on the upper parts. The leaves of the young plant are in a rosette at the base of the stem, as well as along the stem itself. The basal rosette leaves die off as the plant matures. The lower leaves are 5–10 cm long, narrowly oval and broader near the tip than at the base. The upper leaves are narrower, up to 8 cm long. Several flowers are borne singly on each stem branch. The flowers are up to 2 cm across, with five petals, white or pink inside and creamy-yellow on the backs. Each petal is divided deeply into two. The flowers are tightly-rolled by late morning, opening again in the early evening. They are pollinated by nocturnal flying insects. The seed capsule is oval in shape and approximately 15 mm long, opening at the tip. The seeds are dark-brown, kidney-shaped and about 1 mm in diameter.

Similar species: Night-flowering Catchfly is similar to White Campion. The flower of Night-flowering Catchfly has a very distinctive colour and deeply-notched petals. White Campion never has sticky hairs on the upper stem.

Associated uncommon species: Often part of species-rich communities on chalky or sandy soils with species including Rough Poppy, Prickly Poppy, Narrow-fruited Cornsalad, Dense-flowered Fumitory and Corn Chamomile.

HABITAT:

Arable field margins, especially in root-crops.

SOIL TYPE:

Calcareous loams, calcareous sandy loams.

Night-flowering Catchfly: seed capsule × 1

DISTRIBUTION:

Formerly widespread and frequent throughout the south and east of Britain as far north as south east Scotland. It has declined markedly since the 1950s, but is still relatively frequent in East Anglia, and locally on chalk and limestones in central-southern England.

LIFE CYCLE:

Flowers from July to September. Seed is very long-lived. Germination is mainly between March and May but also in the autumn.

J F M A M J J A S O N D

REASONS FOR DECLINE:

Night-flowering Catchfly is highly susceptible to many broad-spectrum herbicides. It may have been affected by the decrease in area of spring-drilled cereals in eastern England, although it grows well in crops such as Sugar Beet.

Small-flowered Catchfly

Silene gallica

IDENTIFICATION:

Small-flowered Catchfly grows to about 30 cm tall with an occasionally-branched stem. The whole plant is hairy, with sticky hairs on the upper parts. The leaves of the young plant are in a rosette at the base of the stem, as well as along the stem itself. The basal rosette leaves die off as the plant matures. The lower leaves are narrowly oval, broader nearer the tip than the base. The upper leaves are narrower and up to 5 cm long. Several flowers are borne singly on each stem branch. The flowers are up to 15 mm across, with five white or pink petals. A variety (var. *quinquevulneraria*) with red-blotched petals has been cultivated as an ornamental plant. The seed capsule is oval in shape and approximately 10 mm long, opening at the tip. The seeds are dark-brown, kidney-shaped and about 0·8 mm in diameter.

Similar species: Small-flowered Catchfly could be confused with White Campion or with small specimens of Night-flowering Catchfly. White Campion does not have a sticky upper stem, Night-flowering Catchfly does not flower during the day. Both these species have larger flowers than Small-flowered Catchfly.

Associated uncommon species: Small-flowered Catchfly is often found in species-rich communities with Weasel's-snout and Corn Marigold. At two sites it occurs with Broad-fruited Cornsalad.

HABITAT:

Arable field margins and, occasionally, dry sunny banks near the sea.

SOIL TYPE:

Sands and sandy loams.

Small-flowered Catchfly: flower × 1

DISTRIBUTION:

This species was formerly widespread throughout England and Wales, particularly in the south. It has now disappeared from much of its former range, with most remaining sites being near the coasts of south-west England and Wales.

LIFE CYCLE:

Flowers from June to October. Seed is thought to be long-lived. Germination in both spring and autumn.

J F M A M J J A S O N D

REASONS FOR DECLINE:

This poorly-competitive species has been badly affected by increases in levels of nitrogen applied and improved crop varieties. Other related species are very susceptible to a wide range of herbicides.

Corn Chamomile
Anthemis arvensis

IDENTIFICATION:
Corn Chamomile is a typical mayweed. The whole plant is covered in short hairs. The leaves are finely divided with narrow, parallel-sided segments, pointed at the tips and have a pleasant, chamomile-like scent. The 'flowers' are actually compound flower-heads made up of numerous small florets and resemble a Daisy. The central florets are yellow, while around the edge are the ray-florets, which have a single long white petal pointing outwards. Among the yellow florets are numerous small chaff-like scales approximately 2·5 mm long and 1 mm wide, tapering to a point. The flowers are larger than those of other mayweeds, and can be up to 4 cm in diameter.

Similar species: Corn Chamomile is similar to Scentless, Scented and Stinking Mayweeds. These differ from Corn Chamomile as follows:

SCENTLESS MAYWEED – Less hairy; no chaffy scales in yellow florets; crushed leaves scentless.

SCENTED MAYWEED – Less hairy; smaller flower-heads; no chaffy scales in yellow florets; crushed leaves similarly scented.

STINKING MAYWEED – Usually less hairy; smaller flower-heads; very narrow chaffy scales in yellow florets; crushed leaves unpleasantly scented.

Associated uncommon species: It usually occurs in species-rich communities with Rough Poppy, Prickly Poppy, Dense-flowered Fumitory, Narrow-fruited Cornsalad and Night-flowering Catchfly.

HABITAT:
Arable field margins on chalky or sandy soils.

SOIL TYPE:
Calcareous sands or chalky loams.

Corn Chamomile: seed (*left*) ×5,
chaffy scale (*centre*) ×5, ray-floret (*right*) ×5

DISTRIBUTION:
The distribution of Corn Chamomile is poorly known due to the difficulty of distinguishing it from other annual mayweeds. Formerly scattered throughout the south and east of England with populations as far north as Inverness in Scotland. It is now present in a few sites on chalk in central-southern England and on sands in Oxfordshire and the East Anglian Breckland.

LIFE CYCLE:
Flowers from June to July.
Seed biology is unknown.
Germination period unknown.

J	F	M	A	M	J	J	A	S	O	N	D
		?								?	

REASONS FOR DECLINE:
In common with other mayweeds, this species is probably susceptible to many herbicides.

Corn Cleavers
Galium tricornutum

IDENTIFICATION:
Corn Cleavers is superficially similar to the commonly occurring Cleavers. It reaches up to 80 cm in length and scrambles among the crop. The leaves are arranged in whorls of 6–9 along the stem. They are up to 4 cm long and 4 mm wide, broader near the tip than at the base and narrowed to a point at the tip. The stems are sharply 4-angled, and the stems and leaves are covered with backward-pointing prickles. The greenish-white flowers are 1·0–1·5 mm in diameter, in groups of three on stems in the leaf axils. The seeds are spherical and approximately 3 mm in diameter, covered in papillae rather than hooked bristles. The seed stalks are strongly recurved.

Similar species: Cleavers and False Cleavers are both similar. The most reliable features that distinguish these species from Corn Cleavers are their straight seed stalks and seeds covered in hooked bristles. Cleavers and False Cleavers also differ in being much more vigorous species, a darker shade of green, and in having leaves that are held at a less acute angle to the stem.

Associated uncommon species:
At the remaining Hertfordshire site, Corn Cleavers grows with Spreading Hedge-parsley, Corn Buttercup, Shepherd's-needle and Prickly Poppy.

HABITAT:
Cereal field margins, road verges.

SOIL TYPE:
Calcareous clay loams.

MANAGEMENT REQUIREMENTS:
Autumn cultivation.

DISTRIBUTION:
Formerly present locally throughout southern England, Corn Cleavers has recently been recorded in only four sites in Britain, in Cambridgeshire, Hertfordshire, Oxfordshire and Essex. It has not been seen in Oxfordshire since 1985.

LIFE CYCLE:
Flowers from June to September. Seed is thought to be very short-lived.
Germination is mainly in the autumn.

J F M A M J J A S O N D

REASONS FOR DECLINE:
Corn Cleavers is a much less competitive species than Cleavers, and has probably been affected by the use of highly competitive crop varieties and high levels of nitrogen applications. The early decline may have been a result of improved seed cleaning technology.

Corn Cleavers: seeds ×5

Corncockle

Agrostemma githago

IDENTIFICATION:

Corncockle can grow taller than 1 m, with flowers at the same height as the cereal. The leaves are long and narrow and not divided, tapering to a point. The whole plant is covered in white hairs that are pressed close to the plant. The flowers are large, up to 3·5 cm across, and trumpet-shaped, borne singly on long stems. They are bright pink with dark streaks, and are surrounded by long, pointed sepals. The black seeds are up to 3·5 mm across.

Similar species: Corncockle is unmistakable.

Associated uncommon species: Formerly in species-rich communities, often associated with Cornflower.

HABITAT:

Formerly on arable land, particularly in Rye crops where it was resown with crop seed each year. Corncockle now occurs occasionally on new road verges and building sites where it does not persist.

SOIL TYPE:

Most frequently on sandy loams, although this may be because this is the preferred soil type for growing Rye.

MANAGEMENT REQUIREMENTS:

Because of its rather special requirements, this species is very difficult to manage for. If any natural arable population should be found, seed should be harvested annually and resown with the crop in a conservation headland.

DISTRIBUTION:

Formerly widespread in southern and eastern England, Corncockle is now extinct apart from occasional appearances after deep-buried seed has been excavated.

LIFE CYCLE:

Flowers from June to August. Seed lives for one or two months after formation, although it can survive much longer if stored dry with crop seed or deeply-buried. Germination is not seasonal.

J F M A M J J A S O N D

REASONS FOR DECLINE:

The main reason for decline was the development of efficient seed-cleaning technology at the end of the 19th century. It is also susceptible to most broad-spectrum herbicides, but can compete with a fertilised crop.

Corncockle: seed pod × 1

Cornflower

Centaurea cyanus

IDENTIFICATION:

Cornflower is a member of the daisy family (Asteraceae). Plants can be up to 1 m tall, usually overtopping the crop, and often with much-branched stems. The leaves are grey-green and hairy. The lower leaves are 10–20 cm long, narrow and variably dissected, the upper leaves are much smaller. The 'flowers' are up to 3 cm in diameter, and are in fact compound heads composed of numerous small individual flowers. These are an intense blue in colour around the edge of the head, with smaller, purplish flowers in the centre. Non-native populations often have plants with white and purple flowers. Seeds are 3 mm long, similar in size to small cereal grains, and have a short, bristly crown of pappus hairs.

Similar species: Cornflower is unlikely to be mistaken, although Field Scabious bears a slight resemblance and can occasionally be found in the edges of arable fields.

Associated uncommon species: Cornflower is often found in species-rich communities with Shepherd's-needle, Spreading Hedge-parsley, Prickly Poppy, Loose Silky-bent and Corn Parsley.

HABITAT:

Arable land, casual on new road verges. Cornflower is frequently a component of 'wildflower mixes'.

SOIL TYPE:

Most frequent on sandy loams, but also on other soils including calcareous clays.

MANAGEMENT REQUIREMENTS:

Autumn cultivation. Although it can also grow well in a spring-sown crop, plants are smaller and produce fewer seed.

Cornflower:
upper leaf (*above*), lower leaf (*below*) × 0·4

DISTRIBUTION:

Once widespread throughout Britain, Cornflower is now only found in a handful of scattered localities in southern and eastern England.

LIFE CYCLE:

Flowers from June to August, and often again in post-harvest stubble.

Seed is sparse and short-lived; deeply-buried seed may survive for longer.

Germination is mainly in the autumn and winter, but some can germinate following spring cultivations.

J F M A M J J A S O N D

REASONS FOR DECLINE:

Once a frequent contaminant of Rye and Flax seed, early 20th century improvements in seed-cleaning technology and the introduction of herbicides created a rapid decline. The use of nitrogen is thought to have had less impact on this species than on many others.

Broad-fruited Cornsalad

Valerianella rimosa

IDENTIFICATION:

Like other cornsalads, Broad-fruited Cornsalad is a slender, often much-branched plant with narrow, spear-shaped leaves, sometimes with a few teeth near the base. This species, like Narrow-fruited Cornsalad, is hairless. The flowers are borne in terminal clusters, but usually there are additional, solitary flowers in the axils of the branches. The five-petalled flowers are white, occasionally tinged with pink, and little more than 2 mm across. The seeds of Broad-fruited Cornsalad are approximately 1·5 mm across, resembling a rounded grape-pip, with a single tooth at the apex. The first seeds are formed by the solitary axillary flowers.

Similar species: All cornsalads are similar, the best way of distinguishing them being their seeds. The most similar species is Narrow-fruited Cornsalad, which has much narrower seeds.

Associated uncommon species: Broad-fruited Cornsalad is normally part of species-rich communities with other rare species including Shepherd's-needle, Spreading Hedge-parsley, Corn Buttercup, Broad-leaved Spurge, Small-flowered Catchfly, Narrow-fruited Cornsalad and Few-flowered Fumitory.

HABITAT:

Arable land.

SOIL TYPE:

Although found mainly on calcareous soils in the past, the remaining sites for this species are on a variety of soil types including calcareous clay and stony, acidic soils.

Broad-fruited Cornsalad: fruit × 12

DISTRIBUTION:

Never a common species; formerly present in sites scattered across southern England and Wales from Cornwall to north Lincolnshire. There may now be as few as ten sites for this species in south-west and central-southern England.

LIFE CYCLE:

Flowers from June to August.
Seed longevity is unknown.
Germination is thought to occur after both autumn and spring cultivations.

J F M A M J J A S O N D

REASONS FOR DECLINE:

Much of the decline in this species has occurred since the 1950s. It is a poorly competitve species, and has been affected by the increased use of nitrogen fertiliser and development of highly competitive crop varieties. It is likely to be susceptible to most broad-spectrum herbicides.

Common Cornsalad

Valerianella locusta

IDENTIFICATION:

Like other cornsalads, Common Cornsalad is a slender, often much-branched plant with narrow, spear-shaped leaves, sometimes with a few teeth near the base. The flowers of Common Cornsalad are borne in dense terminal clusters, but usually there are additional, solitary flowers in the axils of the branches. The five-petalled flowers are symmetrical and pale blue, and little more than 2 mm across. The seeds of Common Cornsalad are formed of two distinct parts with a groove between the two. The whole seed is 1·8–2·5 mm in diameter, and 1·0–1·5 mm wide.

Similar species: All cornsalads are similar, the best way of distinguishing them being their seeds. The most similar species is Keeled-fruited Cornsalad which has much narrower seeds.

Associated uncommon species: Common Cornsalad occurs in a wide range of communities, and is sometimes present where rare species occur.

HABITAT:

Walls, tracks, sand dunes, horticultural land, but only rarely in arable field margins.

SOIL TYPE:

Light calcareous loams, also vestigial soils on walls and tracks.

DISTRIBUTION:

Common Cornsalad is widespread throughout lowland Britain as far north as eastern Scotland and the Hebrides.

LIFE CYCLE:

Flowers from May to June.
Seed is thought to be quite long-lived.
Germination is in the autumn.

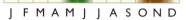

J F M A M J J A S O N D

REASONS FOR DECLINE:

Any decline in this species may be due to the more intensive use of habitats marginal to arable land, and the removal of old walls and tracks. It may have been over-recorded in the past.

Common Cornsalad: seed × 12

Keeled-fruited Cornsalad

Valerianella carinata

IDENTIFICATION:

Like other cornsalads, Keeled-fruited Cornsalad is a slender, often much-branched plant with narrow, spear-shaped leaves, sometimes with a few teeth near the base. The flowers of Keeled-fruited Cornsalad are borne in dense terminal clusters, but usually there are additional, solitary flowers in the axils of the branches. The five-petalled flowers are symmetrical and pale blue, and little more than 2 mm across. The seeds of Keeled-fruited Cornsalad are formed of two distinct parts with a groove between the two. The whole seed is 2·0–2·7 mm in length, and 0·8–1·4 mm wide and thick.

Similar species: All cornsalads are similar, the best way of distinguishing them being their seeds. The most similar species is Common Cornsalad which has much broader seeds.

Associated uncommon species: Keeled-fruited Cornsalad occurs in a wide range of communities, and is sometimes present where rare species occur.

HABITAT:

Walls, tracks, sand dunes, horticultural land, but only rarely arable field margins. A frequent garden weed in south-west England.

SOIL TYPE:

Light calcareous loams, also vestigial soils on walls and tracks.

DISTRIBUTION:

Keeled-fruited Cornsalad is widespread in south and south-west England and Wales below a line running from the Mersey to the Wash, with a highly scattered distribution north of this line.

LIFE CYCLE:

Flowers from May to June.
Seed is thought to be quite long-lived.
Germination is in the autumn.

J F M A M J J A S O N D

REASONS FOR DECLINE:

This species has apparently increased in recent years, although this may be partly as a result of under-recording in the past; other factors may be involved, however.

Keeled-fruited Cornsalad: seed × 12

Narrow-fruited Cornsalad

Valerianella dentata

IDENTIFICATION:

Like other cornsalads, Narrow-fruited Cornsalad is a slender, often much-branched plant with narrow, spear-shaped leaves, sometimes with a few teeth near the base. This species, like Broad-fruited Cornsalad, is hairless. The flowers are borne in terminal clusters, but usually there are additional, solitary flowers in the axils of the branches. The five-petalled flowers are symmetrical, and white, occasionally tinged with pink, and little more than 2 mm across. The seeds of Narrow-fruited Cornsalad are approximately 0·75 mm across with a single tooth at the apex, the first seeds being formed by the solitary axillary flowers.

Similar species: All cornsalads are similar, the best way of distinguishing them being their seeds. The most similar species is Broad-fruited Cornsalad which has much broader, rounded seeds.

Associated uncommon species: Often part of species-rich communities on chalky or sandy soils with species including Rough Poppy, Prickly Poppy, Night-flowering Catchfly, Dense-flowered Fumitory, Corn Parsley and, rarely, Pheasant's-eye, Spreading Hedge-parsley and Broad-fruited Cornsalad.

HABITAT:

Arable field margins particularly in spring crops.

SOIL TYPE:

Light calcareous loams, mainly on chalk, less commonly sandy loams and calcareous clay loams.

DISTRIBUTION:

Formerly widespread and frequent throughout southern and eastern Britain as far north as Scotland. It has declined since the 1940s but still occurs locally in central-southern and eastern England and occasionally on sandy loams elsewhere.

LIFE CYCLE:

Flowers from June to August.
Seed is thought to be quite long-lived.
Germination is mainly in the spring.

J F M A M J J A S O N D

REASONS FOR DECLINE:

Narrow-fruited Cornsalad is thought to be poorly competitive with heavily fertilised modern crop varieties. It is also thought to be susceptible to many broad-spectrum herbicides. As a largely spring-germinating species it has probably been affected by the decreasing area of spring-sown cereal crops.

Narrow-fruited Cornsalad: fruit × 12

Field Cow-wheat

Melampyrum arvense

IDENTIFICATION:

Field Cow-wheat has an erect, branching stem up to 60 cm tall. The glossy-green leaves are spear-shaped and stalkless, up to 8 cm long, often with a few teeth at the base. The flowers are borne in a dense, garishly coloured, cylindrical, terminal spike and have a yellow corolla-tube and two mainly purple-pink lips. Between the flowers are numerous pinkish-red bracts. Each flower forms a capsule containing two seeds. The seeds are similar in appearance and size to a cereal grain. They taste extremely bitter but have a sweet-tasting projection at one end that is thought to be attractive to ants, which take and disperse the seeds. Field Cow-wheat is partially parasitic on a wide range of other flowering plants.

Similar species: The only similar species is Crested Cow-wheat which grows at the edges of woodland. This has a four-sided, rather than cylindrical, flower spike and smaller flowers.

HABITAT:

Open, tussocky grasslands in field boundaries. Formerly in arable field margins, especially on the Isle of Wight where in the 19th century it was locally a serious farming problem.

SOIL TYPE:

Thin soils over chalk; calcareous clays.

MANAGEMENT REQUIREMENTS:

Occasional disturbance of field boundary habitats.

Field Cow-wheat:
upper leaves (*above*),
lower leaves (*below*) × 1

DISTRIBUTION:

Now known from only four locations in southern England (Isle of Wight, Wiltshire and Bedfordshire). It was formerly present in Essex and Norfolk.

LIFE CYCLE:

Flowers from June to September. Seed is very short-lived.

Germination is in the autumn, but plants do not emerge above ground until early spring.

J F M A M J J A S O N D

REASONS FOR DECLINE:

Field Cow-wheat seeds were probably sown with crop seed in the past, and declined with the introduction of efficient seed-cleaning machinery. It may have declined in field boundaries as these have become more efficiently cultivated, removed completely or neglected.

Thale Cress

Arabidopsis thaliana

IDENTIFICATION:
Thale Cress is a slender, erect plant that reaches 35 cm in height, rarely more. It grows from a basal rosette of narrow, usually unlobed, elliptical leaves with short stalks, This rosette forms in the late autumn, survives the winter and dies when the plant begins to flower. There can be several flowering stems, and each of these can be branched with some lanceolate, stalkless leaves. The flowers are in racemes at the ends of the stems. They have short stalks. Each flower has four white petals measuring approximately 3 mm in diameter. The fruit pods are long, thin and curved slightly upwards, measuring 10–18 mm in length by 0·8 mm in diameter.

Similar species: Other small annual species in the cabbage family (Brassicaceae) include Shepherd's-purse and Shepherd's Cress. These can be distinguished from Thale Cress by their flattened triangular- or spoon-shaped fruit capsules respectively. Flixweed is normally a much taller plant with small, yellow flowers and pinnate stem leaves.

Associated uncommon species: Thale Cress is frequently associated with uncommon species of sandy soils, particularly in Breckland.

HABITAT:
Arable field margins, horticultural land, walls, waste ground, tracks.

SOIL TYPE:
Well-drained sandy soils and sandy loams, frequently non-calcareous.

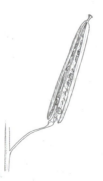

Thale Cress: seed pod ×2

DISTRIBUTION:
Thale Cress is present throughout Britain except north-west Scotland.

LIFE CYCLE:
Flowers from May to June.
Seed can persist in the soil for several years.
Germination is largely in the autumn.

J F M A M J J A S O N D

REASONS FOR DECLINE:
Thale Cress is probably susceptible to many herbicides, and is a non-competitive species that does not grow well in a fully-fertilised modern crop variety.

Broad-leaved Cudweed

Filago pyramidata

IDENTIFICATION:

Broad-leaved Cudweed grows up to 30 cm tall with stems that are usually branched from the base. The whole plant is covered in a felt of grey hairs. The leaves have rounded tips and are broader near the tip than at the base. They are 3–4 mm wide and up to 20 mm long. The tiny flowers of Broad-leaved Cudweed are clustered in dense heads at the apices of the stems. Amongst the flowers are narrow, pointed bracts with yellowish tips. The flower-heads are grey and woolly.

Similar species: Common Cudweed and Red-tipped Cudweed. Common Cudweed has narrow leaves tapering to a point. Red-tipped Cudweed has parallel-sided leaves with pointed tips and red-tipped bracts among the flowers.

Associated uncommon species: On its arable site in Kent it grows with Ground-pine, Rough Poppy, Night-flowering Catchfly and Narrow-fruited Cornsalad. At other sites Wild Candytuft, Fine-leaved Sandwort and Dwarf Mouse-ear are present.

HABITAT:

Margins of unintensively managed arable fields, disturbed calcareous grassland and disturbed and compacted chalk.

SOIL TYPE:

Usually chalk, but sometimes sand, often with very little actual soil, and frequently compacted. In the past, it is thought to have grown more widely on acidic sands and gravels.

MANAGEMENT REQUIREMENTS:

Autumn cultivation is ideal, although plants will also grow after a spring cultivation. Occasional disturbance and scrub removal from grassland sites.

Broad-leaved Cudweed: leaf × 2

DISTRIBUTION:

Broad-leaved Cudweed has been recorded from numerous sites in the eastern half of Britain from Dorset and Kent north to South Yorkshire. It is now known from nine sites in Cambridgeshire, Essex, Kent, Oxfordshire, Surrey and Sussex.

LIFE CYCLE:

Flowers from mid-July to mid-October. Seed is moderately short-lived. Germination is mainly in October and November, with some in March and April.

J F M A M J J A S O N D

REASONS FOR DECLINE:

Arable sites lost by conversion to pasture and overall intensification of management, in particular the use of herbicides, application of high levels of artificial nitrogen to competitive modern crop varieties and the ploughing of post-harvest stubbles.

Narrow-leaved Cudweed

Filago gallica

E

IDENTIFICATION:

Narrow-leaved Cudweed grows up to 25 cm tall, but is often shorter. It can be branched from the base, possibly as a response to grazing by Rabbits. The whole plant is covered in fine, grey hairs. The leaves are long and fine, up to 18 mm long. The flowers of Narrow-leaved Cudweed are very small and are clustered together in dense flower-heads at the apices of the stems. These flower-heads are grey and woolly.

Similar species: Small Cudweed is a similar species of acidic grasslands and heathland. It has much shorter leaves. Other cudweeds of arable and disturbed land have much broader leaves.

Associated uncommon species: Broad-leaved and Red-tipped Cudweeds were formerly present at the last Essex site.

HABITAT:

Arable and horticultural field margins, parched acidic grassland.

SOIL TYPE:

Acidic sands and gravels.

MANAGEMENT REQUIREMENTS:

In arable field margins on nutrient-poor soils, an uncropped strip cultivated biennially in autumn would be suitable.

DISTRIBUTION:

Narrow-leaved Cudweed has been recorded from 30 sites in south-eastern England since 1696. The last native site was near Colchester in Essex, where it was last seen in 1955. It was reintroduced to this last site in 1994 and has also been planted at several sites in Suffolk.

LIFE CYCLE:

Flowers from July to September. Seed longevity is unknown. Germination is mainly in the autumn.

J F M A M J J A S O N D

REASONS FOR DECLINE:

The main reason for its decline in arable habitats was probably change in land use from unintensively managed cultivated land to grassland.

Narrow-leaved Cudweed: leaf ×2

Red-tipped Cudweed
Filago lutescens

IDENTIFICATION:

Red-tipped Cudweed grows up to 25 cm tall, and usually has branched stems. The whole plant is covered in a felt of yellow-tinged grey hairs. The leaves are parallel-sided, 3–4 mm wide and up to 20 mm long with pointed tips. The tiny flowers of Red-tipped Cudweed are clustered in dense flower-heads at the apices of the stems. The flower-heads are grey and woolly. Amongst the flowers are narrow, pointed bracts with dark red tips that fade with age.

Similar species: Common Cudweed, which has narrow leaves tapering to a point; and Broad-leaved Cudweed, which has leaves that are broader near the end than at the base and bracts tipped with yellow-brown rather than red.

Associated uncommon species: Usually found on disturbed grassland sites which are very species-rich, with species including Clustered Clover, Hoary Cinquefoil, Small Cat's-ear and other cudweeds.

HABITAT:

Margins of unintensively managed arable and horticultural fields, disturbed acidic grassland.

SOIL TYPE:

Sand and sandy loams.

MANAGEMENT REQUIREMENTS:

Autumn cultivation. Occasional disturbance and removal of scrub from grassland sites.

Red-tipped Cudweed: leaf × 2

DISTRIBUTION:

Red-tipped Cudweed has been recorded from numerous sites in the eastern half of Britain from Dorset and Kent north to Yorkshire. It is now known from 22 sites: in Surrey, Hampshire, Sussex, Essex, Suffolk, Norfolk and Gloucestershire.

LIFE CYCLE:

Flowers from mid-July to mid-October. Seed is moderately long-lived. Germination is mostly between October and February.

J F M A M J J A S O N D

REASONS FOR DECLINE:

Arable sites have been lost by conversion to pasture and overall intensification of management, in particular the use of herbicides, the application of high levels of artificial nitrogen and the ploughing of post-harvest stubbles.

Cut-leaved Dead-nettle

Lamium hybridum

IDENTIFICATION:

Cut-leaved Dead-nettle grows up to 30 cm tall, with an irregularly-branched stem. The leaves are up to 50 mm long and arranged in pairs on opposite sides of the stem. They are triangular in shape with deep but irregular teeth and with short stalks. Up to 10 flowers are borne in 3–7 whorls at the ends of the stems. The pinkish-purple flowers have bilateral symmetry and are 10–18 mm across with a corolla-tube up to 12 mm long. The flowers often do not open. Each flower produces four seeds.

Similar species: The other dead-nettles are similar. The leaves of Red Dead-nettle are not divided into teeth. The leaves of Henbit Dead-nettle are stalkless, clasping the stem.

Associated uncommon species: Cut-leaved Dead-nettle occurs in a wide range of communities, and is often present where rare species occur.

HABITAT:

Arable and horticultural field margins.

SOIL TYPE:

It can be found on most soil types. It is most common on non-calcareous, sandy loams in eastern England.

DISTRIBUTION:

Throughout England, but more common in the east. It has increased in recent years.

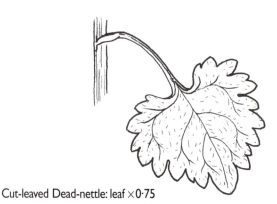

Cut-leaved Dead-nettle: leaf × 0·75

LIFE CYCLE:

Flowers from March to October. Seed longevity is unknown. Germination period is unknown.

J F M A M J J A S O N D

REASONS FOR DECLINE:

Cut-leaved Dead-nettle has become more frequent in recent years.

Henbit Dead-nettle

Lamium amplexicaule

IDENTIFICATION:

Henbit Dead-nettle grows up to 30 cm tall, with an irregularly-branched stem. The leaves are arranged in pairs on opposite sides of the stem. They are round in shape with irregular teeth. The lower leaves have short stalks, the upper leaves are stalkless and clasp the stem. Up to 10 flowers are borne in 3–7 whorls at the ends of the stems. The pinkish-purple flowers are 14–20 mm across and have bilateral symmetry, with a corolla-tube up to 14 mm long. The flowers, particularly those produced early in the summer, often do not open. Each flower produces four seeds.

Similar species: The other dead-nettles are similar, but none have the characteristic leaves clasping the stem.

Associated uncommon species: Henbit Dead-nettle occurs in a wide range of communities, and is often present where rare species occur. It is a typical associate of species-rich communities on chalky soils.

HABITAT:

Arable and horticultural field margins.

SOIL TYPE:

It is most common on well-drained calcareous loams in southern and eastern England, but can be found on a wide range of soil types.

DISTRIBUTION:

Throughout Britain, but more common in the south and east.

LIFE CYCLE:

Flowers from June to August.
Seeds are long-lived.
Germination is mainly in the spring, but continues throughout the summer.

J F M A M J J A S O N D

REASONS FOR DECLINE:

Henbit Dead-nettle is susceptible to many herbicides, and is a non-competitive species that does not grow well in a fully-fertilised modern crop variety. As it also flowers in stubbles it has probably been affected by early ploughing.

Henbit Dead-nettle: flower × 3

Northern Dead-nettle

Lamium confertum

IDENTIFICATION:
Northern Dead-nettle is similar in general form to other dead-nettles. It grows up to 30 cm tall, with an irregularly-branched stem. The leaves have short stalks and are triangular in shape and up to 25 mm long. They are arranged in pairs on opposite sides of the stem but do not clasp the stem. The bilaterally symmetrical flowers are pinkish-purple in colour and 10–18 mm long. They are borne in whorls of 3–7 flowers at the ends of the stems, with each flower surrounded by a calyx of sepals that is as long as the flower. Each flower produces four seeds.

Similar species: The other dead-nettles are similar. However, no other species has the large calyx, and Henbit Dead-nettle can be further distinguished by its upper leaves which clasp the stem.

Associated uncommon species: Northern Dead-nettle occurs in a wide range of communities. It can occur with species such as Corn Marigold, Purple Ramping-fumitory and others.

HABITAT:
Arable and horticultural field margins, other disturbed land.

SOIL TYPE:
It can be found on most soil types within its range.

DISTRIBUTION:
Scotland, Northern Ireland and Isle of Man only. Largely coastal, but also inland in places.

LIFE CYCLE:
Flowers from May to September.
Seed longevity is not known, but may be long.
Germination is likely to be mainly in in the spring.

J F M A M J J A S O N D

REASONS FOR DECLINE:
Declines are likely to be due to the high levels of nitrogen applied to competitive modern crops. The use of broad-spectrum herbicides and increases in the proportion of the cropped area sown with winter cereals.

Northern Dead-nettle: leaf × 0·75

Round-leaved Fluellen

Kickxia spuria

IDENTIFICATION:

Round-leaved Fluellen is a prostrate, creeping plant. Each plant can produce numerous branching stems, sometimes forming mats, particularly in post-harvest stubbles. The leaves are oval in shape, hairy and arranged alternately along the stems. The solitary, bilaterally symmetrical flowers have fused petals that are yellow with a deep purple upper lip and a straight spur. They are borne on stalks in the leaf axils.

Similar species: The only other similar species is the Sharp-leaved Fluellen, from which it can be easily distinguished by the shape of the leaves. Sharp-leaved Fluellen is a less robust plant.

Associated uncommon species: Round-leaved Fluellen is frequently part of arable plant communities including uncommon species on all soil types.

HABITAT:

Arable field margins, open and waste ground.

SOIL TYPE:

Well-drained chalky soils. More frequently on calcareous clays than Sharp-leaved Fluellen, but less frequently on sands.

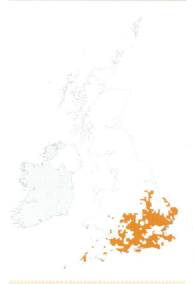

DISTRIBUTION:

Round-leaved Fluellen is present throughout much of lowland England, but is most frequent in the south and east.

LIFE CYCLE:

Flowers from July, lasting until the fields are ploughed or until the plants are killed by frosts.

Seed biology is little-known; seed can probably persist in the soil for several years.

Germination appears to be largely in the spring.

J F M A M J J A S O N D

REASONS FOR DECLINE:

Round-leaved Fluellen is probably susceptible to many herbicides, and is a non-competitive species that does not grow well in a fully-fertilised modern crop variety. Early ploughing of stubbles has also affected this species.

Round-leaved Fluellen: leaf × 1

Sharp-leaved Fluellen

Kickxia elatine

IDENTIFICATION:

Sharp-leaved Fluellen is a prostrate, creeping plant. Each plant can produce numerous branching stems sometimes forming mats, particularly in post-harvest stubbles. The leaves are arranged alternately along the stems. They are triangular in shape, with teeth at the base that project sideways and are hairy. The solitary, bilaterally symmetrical flowers have fused petals that are yellow with a purple upper lip and a straight spur. They are borne on stalks in the leaf axils.

Similar species: The only other similar species is the Round-leaved Fluellen, from which it can be easily distinguished by the shape of the leaves. Round-leaved Fluellen is a more robust plant.

Associated uncommon species: Sharp-leaved Fluellen is frequently part of arable plant communities including uncommon species on all soil types.

HABITAT:
Arable field margins.

SOIL TYPE:
Well-drained sandy and chalky soils. Less frequently on clay.

DISTRIBUTION:
Sharp-leaved Fluellen is present throughout the south of Britain.

LIFE CYCLE:
Flowers from July, lasting until the fields are ploughed or until the plants are killed by frosts.

Seed biology is little-known; seed can probably persist in the soil for several years.

Germination appears to be largely in the spring.

J F M A M J J A S O N D

REASONS FOR DECLINE:
Sharp-leaved Fluellen is probably susceptible to many herbicides, and is a non-competitive species that does not grow well in a fully-fertilised modern crop variety. Early ploughing of stubbles has also affected this species.

Sharp-leaved Fluellen: leaf × 1

Common Fumitory

Fumaria officinalis

IDENTIFICATION (see also page 238):

The leaf segments of Common Fumitory are channelled. The flowers are borne in loose racemes opposed by leaves, at several points along the stem. Two sub-species are recognised: ssp. *officinalis* has 20–60 flowers per raceme, whilst ssp. *wirtgenii* has fewer than 20 flowers per raceme. The flowers are typical of the fumitory family, approximately 6–8 mm long and deep pinkish-red with petals tipped darker red. The lower petal is spatula-shaped, widening just below the tip. On each side of the flower is a white or pink, irregularly toothed sepal approximately up to 1·5 mm wide × 3·5 mm long in ssp. *officinalis* but smaller in ssp. *wirtgenii*. Each flower produces a single fruit which is distinctly wider than long, about 2·5 mm in diameter, those of ssp. *officinalis* being slightly notched at the apex and those of ssp. *wirtgenii* having a small projection.

Similar species: The other fumitories are similar. The five species of ramping-fumitory have larger flowers. Small-flowered, Dense-flowered and Fine-leaved Fumitories have smaller flowers. The fruits of Common Fumitory are very distinctive.

Associated uncommon species: Common Fumitory occurs in a wide range of communities, and is often present where rare species occur.

HABITAT:

Arable and horticultural field margins.

SOIL TYPE:

It can be found on most soil types apart from the most freely-draining and acidic. It is most common on calcareous loams.

DISTRIBUTION:

Throughout Britain, but more common in the south. It is by far the most common fumitory on calcareous and neutral soils although it has declined in recent years.

LIFE CYCLE:

Flowers from mid-June to October. Seed is thought to be long-lived. Germination is mainly in the spring, although some also germinates in the autumn.

J F M A M J J A S O N D

REASONS FOR DECLINE:

Common Fumitory competes poorly with highly fertilised modern crop varieties, and is susceptible to herbicides.

Common Fumitory: fruit (*left*) ×6, flower (*right*) ×3

Dense-flowered Fumitory

Fumaria densiflora

IDENTIFICATION (see also page 238):

The leaf segments of Dense-flowered Fumitory are very narrow and channelled. The flowers are in dense racemes of sometimes as many as 20–25 flowers opposed by leaves, at several points along the stem. The flowers have bilateral symmetry, with the petals fused to form a tube, but separating towards the tip. They are small compared with other fumitories, approximately 6–7 mm long, and are pink/purple with black tipped petals. On each side of the flower is a large, white, irregularly toothed sepal of approximately 2 mm × 3 mm. Each flower produces a single spherical fruit about 2 mm in diameter, rounded at the apex.

Similar species: Few-flowered Fumitory and Fine-leaved Fumitory are similar but have much smaller sepals.

Associated uncommon species: It typically occurs in species-rich communities with Rough Poppy, Prickly Poppy, Narrow-fruited Cornsalad, and Field Gromwell. At some sites Spreading Hedge-parsley, Shepherd's-needle, Red Hemp-nettle, Few-flowered Fumitory and Small-flowered Fumitory occur.

HABITAT:

Arable land and occasionally on coastal sands, calcareous clay and sandy loams.

SOIL TYPE:

It is largely confined to arable land on light calcareous soils, mainly on chalk, although occasionally found on coastal sands, calcareous clay and sandy loams.

MANAGEMENT REQUIREMENTS:

Spring cultivation.

Dense-flowered Fumitory: fruit (*left*) ×6, flower (*right*) ×3

DISTRIBUTION:

Dense-flowered Fumitory still occurs widely on the chalk of southern and south-eastern England with a second centre of distribution in south-east Scotland. There are isolated sites elsewhere in southern England.

LIFE CYCLE:

Flowers from mid-June to August.
Seed is long-lived.
Germination is mainly in the spring, some in autumn.

J F M A M J J A S O N D

REASONS FOR DECLINE:

Fumitories are susceptible to a wide range of broad-spectrum herbicides. Dense-flowered Fumitory is a poorly competitive species and has probably been adversely affected by the increased use of nitrogen and the arrival of competitive crop varieties. The change from spring Barley to winter cereals is also a likely factor.

Few-flowered Fumitory

Fumaria vaillantii

IDENTIFICATION (see also page 238):

The typically blue-green leaf segments of Few-flowered Fumitory are relatively flat compared to other small-flowered fumitories. The flowers are in loose racemes of usually 6-16 flowers, sometimes more, opposed by leaves, at several points along the stem. The flowers have bilateral symmetry, with the petals fused to form a tube, but separating towards the tip. They are very small compared with other fumitories, about 5–6 mm long, and are pale pink with the lateral petals tipped blackish-red. On each side of the flower is a very small toothed, pale purple sepal, approximately 0·5 mm × 1·0 mm. Each flower produces a single spherical fruit about 2 mm in diameter.

Similar species: Fine-leaved and Dense-flowered Fumitories are similar. The leaf segments of Fine-leaved Fumitory are much narrower and channelled, and the flowers are white. Dense-flowered Fumitory has very distinctive large, white sepals.

Associated uncommon species: It usually grows in species-rich communities with other uncommon species including Dense-flowered and Fine-leaved Fumitories, Prickly Poppy, Rough Poppy and Narrow-fruited Cornsalad.

HABITAT:

Arable field margins.

SOIL TYPE:

Calcareous loams.

MANAGEMENT REQUIREMENTS:

Spring cultivation.

DISTRIBUTION:
Few-flowered Fumitory has always been more or less confined to the chalk of southern and eastern England.

LIFE CYCLE:
Flowers from mid-June to early August.
Seed is probably quite long-lived.
Germination is mainly in the spring.

J F M A M J J A S O N D

REASONS FOR DECLINE:
Few-flowered Fumitory competes poorly with highly fertilised modern crop varieties, and is susceptible to herbicides.

Few-flowered Fumitory: fruit (*left*) ×6, flower (*right*) ×3

Fine-leaved Fumitory

Fumaria parviflora

Nationally Scarce

IDENTIFICATION (see also page 238):

The leaves of Fine-leaved Fumitory, like all fumitory species, are blue-green and are irregularly divided almost to the leaf midrib into narrowly-lobed segments. In Fine-leaved Fumitory these segments are very narrow and channelled. The flowers are in dense racemes of up to 20 flowers opposed by leaves, at several points along the stem. The flowers have bilateral symmetry, with the petals fused to form a tube, but separating towards the tip. They are very small compared with other fumitories, approximately 5–6 mm long, white with petals tipped blackish-red. On each side of the flower is a small white, irregularly toothed sepal of approximately 0·8 mm × 1·0 mm. Each flower produces a single spherical fruit approximately 2 mm in diameter, sometimes with a short beak at the apex.

Similar species: Few-flowered Fumitory and Dense-flowered Fumitory are similar. The leaf-segments of Few-flowered Fumitory are much broader, and the flowers are pink. Dense-flowered Fumitory has very distinctive large, white sepals.

Associated uncommon species: It usually grows in species-rich communities with other uncommon species including Dense-flowered and Few-flowered Fumitories, Prickly Poppy, Rough Poppy and Narrow-fruited Cornsalad.

HABITAT:
Arable field margins.

SOIL TYPE:
Calcareous loams.

MANAGEMENT REQUIREMENTS:
Spring cultivation.

Fine-leaved Fumitory: fruit (*left*) ×6, flower (*right*) ×3

DISTRIBUTION:
Fine-leaved Fumitory has always been more or less confined to the chalk of southern and eastern England.

LIFE CYCLE:
Flowers from mid-June to early August.
Seed is thought to be long-lived.
Germination is mainly in the spring.

J F M A M J J A S O N D

REASONS FOR DECLINE:
Fine-leaved Fumitory competes poorly with highly fertilised modern crop varieties, and is susceptible to herbicides.

Common Ramping-fumitory

Fumaria muralis ssp. *boraei*

IDENTIFICATION (see also page 239):

The leaf segments of Common Ramping-fumitory are relatively broad and flat. The flowers are in loose racemes of 12–15 flowers opposed by leaves, at several points along the stem. These racemes are shorter than their stalks. The flowers have bilateral symmetry, with the petals fused to form a tube, but separating towards the tip. They are 9–11 mm long, and are pinkish-red with petals tipped blackish-red. On each side of the flower is a pale sepal, approximately 1·5–3·0 mm × 3·0–5·0 mm and often toothed all round. Each flower produces a single, more or less spherical, fruit about 2·5 mm in diameter, with a rounded apex.

Similar species: The other ramping-fumitories are very similar. Tall Ramping-fumitory has paler, salmon-pink flowers. Martin's Ramping-fumitory has racemes longer than their stalks. Purple Ramping-fumitory has recurved fruit stalks. Western and White Ramping-fumitories have larger, paler flowers. Common, Small-flowered, Fine-leaved and Dense-flowered Fumitories have smaller flowers.

Associated uncommon species: Common Ramping-fumitory occurs in a wide range of communities, and is often present where rare species occur.

HABITAT:

Arable and horticultural field margins; hedge banks.

SOIL TYPE:

It is found mainly on more freely-draining and acidic loams.

DISTRIBUTION:

Throughout Britain, but more common in the west. It is the most common fumitory on acidic loamy soils in the west of Britain although it has declined in recent years.

LIFE CYCLE:

Flowers from mid-June to October. Seed is thought to be long-lived. Germination is mainly in the spring, although there is some in the autumn.

J F M A M J J A S O N D

REASONS FOR DECLINE:

Common Ramping-fumitory competes poorly with highly fertilised modern crop varieties, and is susceptible to herbicides.

Common Ramping-fumitory: fruit (*left*) ×6, flower (*right*) ×3

Martin's Ramping-fumitory

Fumaria reuteri

IDENTIFICATION (see also page 239):

In common with other fumitories with large flowers the leaf segments of Martin's Ramping-fumitory are relatively broad and flat. The flowers are in racemes of 15–20 flowers opposed by leaves, at several points along the stem. The raceme is longer than its stalk. The flowers have bilateral symmetry, with the petals fused to form a tube, but separating towards the tip. They are approximately 11–13 mm long and pink with petals tipped blackish-red. On each side of the flower is a more or less untoothed sepal, approximately 2.5×4.0 mm. Each flower produces a single spherical fruit about 2.5 mm in diameter.

Similar species: Other fumitories, especially Tall Ramping-fumitory, which has smaller flowers and sepals that are more toothed, and Common Ramping-fumitory, which has racemes shorter than their stalks. Few-flowered, Dense-flowered and Fine-leaved Fumitories have smaller flowers. The other fumitories with large flowers have racemes the same length or shorter than their stalks.

Associated uncommon species: On the Isle of Wight it occurs with Purple Ramping-fumitory and Tall Ramping-fumitory.

HABITAT:

Arable fields, hedge bottoms and allotments.

SOIL TYPE:

Sandy loams.

MANAGEMENT REQUIREMENTS:

Spring cultivation.

DISTRIBUTION:

Martin's Ramping-fumitory has only ever been recorded from 14 sites in southern England. Recent records come from three sites in a small area near Truro, Cornwall and from one site on the Isle of Wight, where it grows on allotments.

LIFE CYCLE:

Flowers from mid-June to early August. Seed is probably quite long-lived. Germination is mainly in the spring but can occur in the summer and autumn.

J F M A M J J A S O N D

REASONS FOR DECLINE:

One of the main reasons for the decline of this species is the reversion of arable and horticultural land to grassland. It is susceptible to herbicides and poorly competitive with highly fertilised modern crop varieties.

Martin's Ramping-fumitory: fruit (*left*) ×6, flower (*right*) ×3

Purple Ramping-fumitory

Fumaria purpurea

IDENTIFICATION (see also page 239):

In common with other large-flowered fumitories, the leaf segments of Purple Ramping-fumitory are relatively broad and flat. The flowers are in racemes of 20–25 flowers opposed by leaves, at several points along the stem. The raceme is about the same length as its stalk, but sometimes shorter. The typical fumitory flowers are 10–13 mm long, pinkish-purple with the petals tipped darker purple When viewed from the side the wings of the upper petal hide its keel. On each side of the flower is a large, 6 mm × 3 mm, oblong-shaped pale sepal that is, apart from some teeth at the base, more or less untoothed. These sepals are half the length of the flower. Each flower produces a single squarish fruit about 2·5 mm in diameter. As the fruit ripens the fruit stalks (although not all) can become variably but gently-recurved.

Similar species: Other fumitories. Few-flowered, Dense-flowered and Fine-leaved Fumitories have smaller flowers. Whilst other larger-flowered species may have recurved fruit stalks they have fewer flowers per raceme. White Ramping-funitory has a similar number of flowers per raceme but differs in its fruit stalks that are rigidly recurved.

HABITAT:
Arable and horticultural field margins, Cornish hedges.

SOIL TYPE:
Sandy loams.

MANAGEMENT REQUIREMENTS:
Spring cultivation.

Purple Ramping-fumitory: fruit (*left*) ×6, flower (*right*) ×3

DISTRIBUTION:
Found mainly in coastal regions of south-west Britain from Hampshire north to Orkney. It is most frequent in Cornwall and Lancashire. This species is endemic to Britain and Ireland.

LIFE CYCLE:
Flowers from mid-June to early August.
Seed is probably quite long-lived.
Germination is mainly in the spring but can occur in the summer.

J F M A M J J A S O N D

REASONS FOR DECLINE:
One of the main reasons for the decline of this species in the areas in which it occurs is the conversion of arable and horticultural land to pasture. It is susceptible to herbicides and poorly competitive with highly fertilised modern crop varieties.

Tall Ramping-fumitory

Fumaria bastardii

IDENTIFICATION (see also page 239):
Like all fumitories, Tall Ramping-fumitory is a scrambling, much-branched plant. The leaves, as in the other fumitory species, are blue-green and are irregularly divided almost to the leaf midrib into narrow lobes. The leaf segments of Tall Ramping-fumitory are relatively broad and flat. The flowers are borne in racemes of 10–18 flowers opposed by leaves, at several points along the stem. The raceme is longer than its stalk. The flowers have bilateral symmetry, with the petals fused to form a tube, but separating towards the tip. They are approximately 9–11 mm long, and are pale salmon-pink with only the lateral petal tipped darker purple. On each side of the flower is a small, white, toothed sepal about 3 mm long × 1·5 mm wide. The bract is less than half the length of the flower stalk. Each flower produces a single spherical fruit about 2·0–2·4 mm in diameter which is wrinkly when dry. The fruit stalks are never recurved.

Similar species: Other fumitories. Martin's Ramping-fumitory has larger sepals with fewer teeth, The other Ramping-fumitories have racemes the same length or shorter than their stalks and larger sepals. Few-flowered, Dense-flowered and Fine-leaved Fumitories have smaller flowers.

HABITAT:
Arable and horticultural field margins; Cornish hedges.

SOIL TYPE:
Sandy loams.

DISTRIBUTION:
Found mainly in coastal regions of south-west Britain from Hampshire north to the Outer Hebrides. It is most frequent in Cornwall and west Wales.

LIFE CYCLE:
Flowers from mid-June to early August.
Seed is probably quite long-lived.
Germination is mainly in the early spring.

J F M A M J J A S O N D

REASONS FOR DECLINE:
One of the main reasons for the decline of this species in the areas in which it occurs is the conversion of arable and horticultural land to pasture. It is susceptible to herbicides and poorly competitive with highly fertilised modern crop varieties.

Tall Ramping-fumitory: fruit (*left*) ×6, flower (*right*) ×3

Western Ramping-fumitory
Fumaria occidentalis

IDENTIFICATION (see also page 239):

Ramping-fumitories are scrambling, much-branched plants, and this species is the most vigorous. The leaves are blue-green and are irregularly divided almost to the leaf midrib into narrow lobes. The leaf segments of Western Ramping-fumitory are relatively broad and flat. The flowers are in loose racemes of 12–20 flowers opposed by leaves, at several points along the stem. The raceme is about the same length as its stalk, usually shorter. The flower petals are fused to form a tube, separating towards the tip. They are larger than in any other fumitory, 12–15 mm long. The flowers are at first white, but soon after flowering the wings of the upper petal turn pink with a distinctive white margin. As the flower matures the area of pink coloration increases and the white margin is lost. The lower petal has broad spreading edges and the lateral petals are tipped dark purple-red. On each side of the flower is a sepal, up to 3·5 mm wide × 5·5 mm long, with a few teeth which seems rather small in proportion to the rest of the flower. Each flower produces a single fruit about 3 mm in diameter with a small projection at the end.

Similar species: Other fumitories, although Western Ramping-fumitory has very distinctive fetaures: its large flower; large fruit and the presence of white margins to the wings of upper petals in young flowers.

HABITAT:

Arable and horticultural field margins; Cornish hedges.

SOIL TYPE:

Sandy loams.

DISTRIBUTION:
Western Ramping-fumitory is endemic to Cornwall and the Isles of Scilly.

LIFE CYCLE:
Flowers from late May to October. Seed is probably quite long-lived. Germination is probably mainly in the autumn.

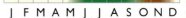

J F M A M J J A S O N D

REASONS FOR DECLINE:
One of the main reasons for the decline of this species in the area in which it occurs is the conversion of arable and horticultural land to pasture. It is susceptible to herbicides and poorly competitive with highly fertilised modern crop varieties.

Western Ramping-Fumitory: fruit (*left*) ×6, flower (*right*) ×3

Cut-leaved Germander

Teucrium botrys

IDENTIFICATION:

Cut-leaved Germander is a member of the mint family (Lamiaceae). It has a much-branched, erect, leafy stem, up to 30 cm in height. The whole plant is hairy. The leaves are up to 2·5 cm long and oval in shape, deeply divided into parallel-sided segments, 1–2 mm wide. The flowers occur in whorls along the stems in the axils of the leaves. The flowers are pinkish-red, with very small upper lips and a long lower lip 8 mm long with a tube 6 mm long enclosed within the calyx.

Similar species: Cut-leaved Germander is unmistakable.

Associated uncommon species: It occurs with Ground-pine at two sites and Red Hemp-nettle at one.

HABITAT:

Exposed chalk cliffs and spoil; irregularly cultivated arable field margins and fallow.

SOIL TYPE:

Thin soils over chalk and Jurassic limestone.

MANAGEMENT REQUIREMENTS:

Maintenance of soil disturbance.

DISTRIBUTION:

Cut-leaved Germander has always been rare and confined to a few sites in the south of England. It now occurs in only six localities in Surrey, Kent, Hampshire and Gloucestershire.

LIFE CYCLE:

Flowers from July to September. Seed longevity is unknown. Germination period is unknown.

J	F	M	A	M	J	J	A	S	O	N	D
?									?		

REASONS FOR DECLINE:

This may have been a species that relied at least partially on introduction with small-seeded crops. It has been affected by intensive cultivation of field-edge habitats, increases in the use of nitrogen and the development of more competitive crop varieties.

Cut-leaved Germander: leaf × 2

Grass-poly

Lythrum hyssopifolium

IDENTIFICATION:

Grass-poly is a low-growing plant up to 20 cm tall, with a branched stem that sometimes creeps along the ground. The leaves vary in shape depending on their position on the plant. Those near the stem-base are oval, whereas those higher up the stem are narrower. The uppermost leaves are very narrow and parallel-sided. The singular flowers are up to 5 mm in diameter, in the axils of the leaves. The petals are pink.

Similar species: Grass-poly can be confused with the much more widespread Knotgrass, the leaves of which are blue-green with white, papery stipules sheathing the stem at the bases of the leaves. Flowers are usually 2–3 in each leaf axil.

Associated uncommon species: Broad-leaved Cudweed formerly grew at the Cambridgeshire site, at which several uncommon mosses and liverworts still also occur.

HABITAT:

Grass-poly has ecological requirements that are very specialised. It grows in areas that are wet in the winter but which dry out in the spring. On arable land these areas usually remain crop-free, as the autumn-sown crop dies in the waterlogged conditions.

SOIL TYPE:
Water-retentive.

MANAGEMENT REQUIREMENTS:
Seasonally wet areas should continue being cultivated.

DISTRIBUTION:

Formerly scattered throughout southern England, particularly around London, Cambridgeshire and Oxfordshire. Grass-poly is now known from only five sites, the three arable sites being in Cambridgeshire, Oxfordshire and Dorset.

LIFE CYCLE:

Flowers from June to July.
Seed is probably quite long-lived.
Germination is in the spring after the ground dries out.

J F M A M J J A S O N D

REASONS FOR DECLINE:
Field drainage and herbicide use.

Grass-poly: seed pod ×6

Field Gromwell

Lithospermum arvense

IDENTIFICATION:

Field Gromwell is normally an erect plant growing up to 60 cm tall, but can sometimes be branched and scarmbling, growing up to 1·5 m in length. The whole plant is bristly. The leaves are narrow, up to 6 cm long and broader near the tip than at the base. The flowers, which are up to 4 mm in diameter, have white petals and are formed both in the leaf axils and on short stems near the apex of the plant. Each flower produces four hard, warty seeds about the size and shape of a grape-pip.

Similar species: Common Gromwell is a similar species of woodland edges and hedgerows. It differs in being a perennial, and in its leaves having prominent lateral veins.

Associated uncommon species: On chalk and limestone soils in the south of England it frequently occurs in species-rich communities with Rough Poppy, Prickly Poppy, Narrow-fruited Cornsalad, Shepherd's-needle and Dense-flowered Fumitory.

HABITAT:

Arable field margins.

SOIL TYPE:

Calcareous loams, calcareous clay loams.

DISTRIBUTION:

Field Gromwell was formerly widely distributed throughout central-southern and eastern England and the Midlands, north to the Humber. It is now much less common within this range, and has gone from many of its northern and western sites. It can still be very locally frequent, and in one part of the south Midlands it has recently caused problems for farmers.

LIFE CYCLE:

Flowers from mid-May to July. Seed is quite long-lived. Germination occurs both between October and December and in March and April.

J F M A M J J A S O N D

REASONS FOR DECLINE:

Although Field Gromwell competes well with cereal crops, it is susceptible to many herbicides.

Field Gromwell:
seed cluster in leaf axil (*left*) ×6, seed (*right*) ×6

Ground-pine

Ajuga chamaepitys

IDENTIFICATION:

Ground-pine resembles a pine seedling in both look and smell. The whole plant is hairy. It grows to a maximum of 20 cm tall, often branching from the base. The leaves are up to 4 cm long, divided into three linear lobes. The flowers are borne in the leaf axils in pairs at each node. The flowers have bilateral symmetry with a very short upper lip and a three-lobed lower lip. The petals are yellow with red spots on the lower lip.

Similar species: Ground-pine is unmistakable.

Associated uncommon species: Other field edge species including Cut-leaved Germander, Broad-leaved Cudweed and Rough Marsh-mallow, and arable field species including Rough Poppy, Prickly Poppy and Narrow-fruited Cornsalad.

HABITAT:

The occasionally disturbed edges of arable fields, fallow and set-aside land and rabbit-disturbed grassland, usually on south-facing sunny slopes.

SOIL TYPE:

Thin soils over chalk.

MANAGEMENT REQUIREMENTS:

Occasional cultivation of field edges where it occurs in such habitats. Occasional fallowing of arable fields.

DISTRIBUTION:

Ground-pine has always been a local species of the North Downs, Hampshire and the East Anglian chalk. It is now restricted to a few sites in Kent, Surrey, Hampshire and Bedfordshire.

LIFE CYCLE:

Flowers from May to October, often in stubble.
Seed is thought to be very long-lived. Germination is either in autumn or spring.

J F M A M J J A S O N D

REASONS FOR DECLINE:

Intensive cultivation of field edges, use of large quantities of nitrogen fertilisers, competitive crop varieties and the application of broad-spectrum herbicides. Early ploughing of post-harvest stubbles.

Ground-pine fruit: (*left*) ×6, seed (*right*) ×6

Knotted Hedge-parsley

Torilis nodosa

IDENTIFICATION:

Knotted Hedge-parsley is a rather unusual member of the carrot family (Apiaceae). It has a spreading stem, sometimes supported by crop plants, and grows up to 50 cm in length. The leaves are 2-pinnate, in rosettes at the base of the plant but also along the stem. The flowers are small and pinkish-white, about 1 mm in diameter, and borne in dense, stalkless clusters, 5–10 mm across, along the stem, opposite the leaves. The seeds are oval and 2·5–3·5 mm in diameter with long or short spines.

Similar species: Several other annual and biennial species in the carrot family – Shepherd's-needle, Fool's Parsley, Wild Carrot, Upright Hedge-parsley, Spreading Hedge-parsley, Cow Parsley – occur in similar habitats. The flower-clusters of Knotted Hedge-parsley are characteristic.

Associated uncommon species: On arable land, Knotted Hedge-parsley usually occurs in species-rich communities with uncommon species including Corn Parsley, Small-flowered Buttercup, Rough Poppy and Prickly Poppy.

HABITAT:

Open, parched grasslands near the sea on cliff-tops and sea-walls and sometimes inland on limestones; arable field margins.

SOIL TYPE:

Calcareous clay loams and sometimes lighter soils.

DISTRIBUTION:

Mainly in the south and east of England north to East Yorkshire. In coastal habitats in the south-west of England and Wales and along the north-east coast of England.

LIFE CYCLE:

Flowers from May to July.
Seed longevity is unknown.
Germination is mainly in the autumn.

J F M A M J J A S O N D

REASONS FOR DECLINE:

This may have been a species that relied at least partially on introduction with small-seeded crops. It has been affected by intensive cultivation of field-edge habitats, increases in the use of nitrogen fertilisers and the development of more competitive crop varieties.

Knotted Hedge-parsley: fruit ×5

Spreading Hedge-parsley

Torilis arvensis

IDENTIFICATION:

Spreading Hedge-parsley is a member of the carrot family (Apiaceae). It grows up to 40 cm tall and is often much-branched. The whole plant is covered in short bristles, giving a frosted appearance. The leaves are triangular in overall shape, and are usually 2-pinnately divided. The flowers are borne in umbrella-shaped clusters that each have 3–5 rays and are 10–25 mm across. Each flower is small, about 2 mm in diameter, with white or pinkish unequally-sized petals. There are several simple, linear bracts at the base of each part of the flower cluster. The seeds are oval, between 4–6 mm long and are covered in long, hooked spines which help the seeds cling to fur and clothing.

Similar species: Several other annual and biennial umbellifers are similar: Shepherd's-needle, Fool's Parsley, Wild Carrot, Upright Hedge-parsley, Cow Parsley. However, the fruit of Spreading Hedge-parsley is characteristic.

Associated uncommon species: Spreading Hedge-parsley normally occurs in species-rich communities with other rare species including Shepherd's-needle, Corn Buttercup, Broad-leaved Spurge and Corn Parsley.

HABITAT:
Arable fields.

SOIL TYPE:
Clay loams and calcareous clay loams.

MANAGEMENT REQUIREMENTS:
Autumn cultivation.

DISTRIBUTION:

Formerly widespread in southern and eastern England as far north as Yorkshire. It has declined dramatically since the 1940s, and now occurs in isolated sites from Devon to Suffolk. It is probably now most frequent in Somerset and East Anglia.

LIFE CYCLE:

Flowers from July to September. Seed is thought to be short-lived. Germination is almost entirely between October and December.

J F M A M J J A S O N D

REASONS FOR DECLINE:

Increase in nitrogen applications to competitive modern crop types. Its susceptibility to herbicides is unknown. It produces fruit very late in the summer and may have been disadvantaged by earlier harvesting dates (especially for winter Barley) and early ploughing of stubbles.

Spreading Hedge-parsley: fruit ×5

Downy Hemp-nettle
Galeopsis segetum

IDENTIFICATION:
Downy Hemp-nettle is a member of the mint family (Lamiaceae) and grows up to 50 cm tall. The stems are not thickened at the leaf nodes. The velvety leaves are 1·5–8·0 cm in length and up to 3 cm wide. They are narrowly spear-shaped, with 3–9 pairs of teeth and a short stalk. The flowers are up to 2·5 cm in length and are typical of this family, having bilateral symmetry with a large lower lip and a very long corolla-tube, four times as long as the calyx. The petals are sulphur-yellow. Each flower produces four seeds.

Similar species: The hemp-nettles can be distinguished by flower colour. Common Hemp-nettle has purple, pink or white flowers and Red Hemp-nettle has pinkish-red flowers. Large-flowered Hemp-nettle has slightly larger flowers, with the corolla-tube twice as long as the calyx and the lower lip usually tinged with violet. In addition, the stems are thickened at the leaf nodes.

Associated uncommon species: In central Europe, Downy Hemp-nettle is often associated with very species-rich communities including Fingered Speedwell, Lamb's Succory and Shepherd's Cress.

HABITAT:
Arable field margins.

SOIL TYPE:
Sandy and stony loams.

E?

DISTRIBUTION:
Historically this species was found in several places in central England. In recent years it has only been seen in one site in north Wales, where it was last recorded in 1975.

LIFE CYCLE:
Flowers from July to August.
Seed is thought to be long-lived.
Germination is thought to be mainly in the spring.

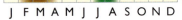

J F M A M J J A S O N D

REASONS FOR DECLINE:
Reasons for decline are unknown, although the last known site has been under pasture since the last record in 1975.

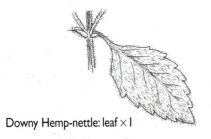

Downy Hemp-nettle: leaf × 1

Red Hemp-nettle

Galeopsis angustifolia

IDENTIFICATION:

Red Hemp-nettle is a relatively low-growing plant with an erect stem up to 50 cm, staying well below the crop canopy. On very nutrient-poor soils plants can be as little as 1 cm tall with a single flower. The leaves are 1·5–8·0 cm in length and less than 1 cm wide with a short stalk. They are narrowly spear-shaped with a serrated edge and lightly covered wth soft hairs. The flower has bilateral symmetry, with a large lower lip and very long corolla-tube. The pinkish-red petals have white patches at their bases. Each flower produces four seeds.

Similar species: This species is unmistakable when flowering; it could possibly be mistaken for Red Bartsia when not in flower.

Associated uncommon species: Red Hemp-nettle occurs typically in species-rich communities that include Rough and Prickly Poppies, Field Gromwell, Narrow-fruited Cornsalad, and Dense-flowered Fumitory. It is also present at sites with Few-flowered and Small-flowered Fumitories, Shepherd's-needle and Spreading Hedge-parsley.

HABITAT:

Arable fields, coastal shingle in the south-east of England and limestone scree elsewhere. On Salisbury Plain, large numbers grow in tank tracks across former arable land.

SOIL TYPE:

As an arable plant, Red Hemp-nettle is confined to light, chalky soils.

MANAGEMENT REQUIREMENTS:

Spring cultivation.
Stubbles should be left after
harvest in the autumn.

DISTRIBUTION:

Once widespread on calcareous soils from Devon to Yorkshire, it has declined since the 1950s and is now restricted to south and south-east England from Dorset to Cambs.

LIFE CYCLE:

Flowers from July to October; plants often regrow in stubbles after harvest.
Seed longevity is unknown, but it is likely to be long-lived.
Germination appears to occur entirely in the spring.

J F M A M J J A S O N D

REASONS FOR DECLINE:

Susceptible to many broad-spectrum herbicides. The increase in nitrogen application and the development of nitrogen-demanding crop varieties are likely to have played a role. A spring-germinating species, the tendency to replace spring Barley with winter Barley is also likely to be a factor.

Red Hemp-nettle: leaf × 1

Cornfield Knotgrass

Polygonum rurivagum

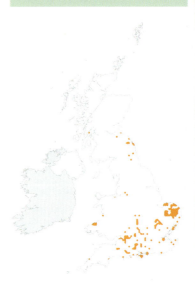

IDENTIFICATION:

Although very similar to the common Knotgrass, this species rarely grows to more than 30 cm tall, and is more erect in growth habit. The stems are usually branched. The leaves are relatively long and thin, 15–35 mm long by 2–4 mm wide, with acute tips. Leaves on the main stem are larger than those on the branches. The leaves and stem are hairless and grey-green in colour. At the base of the leaves are white, papery ochreae, usually *ca.* 10 mm long, that form a tube around the stem. These ochreae are brownish-red below. Flowers are borne singly or in pairs in the leaf axils. They have pinkish petals, and the fruit, measuring approximately 3·0 mm × 1·5 mm, projects slightly from the flower.

Similar species: Other annual knotgrasses, in particular Knotgrass, which is best distinguished by its size: up to 200 cm long; much broader, 5–15 mm wide leaves; shorter, silver-white ochreae, about 5 mm long; and fruit that is enclosed within the flower. The similar Northern Knotgrass occurs in northern areas and differs in having broader, 5–18 mm wide, leaves; leaf stalks that project from the ochreae; and larger fruit, approximately 4·0 mm × 2·5 mm. Equal-leaved Knotgrass has 2–3 flowers in the leaf axils and rounder leaves, about 20 mm long by 5 mm wide

Associated uncommon species: On chalky soils in the south of England, Cornfield Knotgrass often grows in a species-rich community that includes Rough and Prickly Poppies, Grass-poly, Red Hemp-nettle, Narrow-fruited Cornsalad, Night-flowering Catchfly and uncommon fumitories.

SOIL TYPE:

Mainly on light chalky loams, but also on calcareous clays and sandy loams.

Cornfield Knotgrass: fruit (*left*) × 6, seed (*right*) × 6

DISTRIBUTION:

Cornfield Knotgrass has been recorded largely from the chalk areas of central-southern and south-eastern England and East Anglia with several records from the West Midlands, Wales and further north.

LIFE CYCLE:

Flowers from July to October.
Seed is very long-lived.
Germination is thought to be mainly in the spring.

J F M A M J J A S O N D

REASONS FOR DECLINE:

The past distribution of this species is poorly known. This species is likely to have become less common as a result of increased use of herbicides and nitrogenous fertilisers and more competitive crop varieties. Changes from spring to winter drilling of cereals will also have disdvantaged this species.

Larkspur
Consolida ajacis

IDENTIFICATION:
Larkspur is an erect plant 25–60 cm in height. The hairless leaves are finely divided into long, narrow segments. Larkspur has large, bilaterally symmetrical flowers that are deep indigo-blue in colour and borne singly on the end of short stem branches. The flowers are 2–5 mm in diameter, much smaller than the sepals, and with a slender spur. Populations derived from garden plants often have white and pink flowers. The seed pods are up to 25 mm long and covered with downy hairs.

Similar species: The only similar species are other, rare species of larkspur, the Eastern Larkspur and the Forking Larkspur, both of which have hairless seed pods.

HABITAT:
Arable field margins; as a casual on rubbish tips and waste ground.

SOIL TYPE:
On sandy and calcareous loams.

DISTRIBUTION:
The distribution of this species is poorly known. In the past, Larkspur was found most frequently in East Anglia, and still occasionally occurs in this area.

LIFE CYCLE:
Flowers from June to July.
Seed longevity unknown.
Germination period is unknown.

J	F	M	A	M	J	J	A	S	O	N	D
?											?

REASONS FOR DECLINE:
Unknown. This species largely disappeared before the introduction of herbicides and highly competitive crop varieties, but the seeds are not typical of those of species that would have been harvested and resown with the crop.

Larkspur: seed case (*left*) ×2, seed (*right*) ×10

Field Madder
Sherardia arvensis

IDENTIFICATION:
Field Madder resembles a small version of the commonly occurring Cleavers. It is a mat-forming plant with spreading, branching stems. The leaves are borne along the stems in whorls. These whorls have six leaves near the ends of the stems, but only four near the bases. The lower leaves are broadly spear-shaped, whereas the upper leaves are narrow. They are 5–15 mm long and have prickly edges. The lilac flowers, each 2–3 mm across, are in dense, short-stalked clusters of 4–10 at the ends of stems and in leaf axils.

Similar species: Before flowering, Field Madder resembles a small, prostrate cleavers. However the four-leaved whorls of the lower leaves are very distinctive. The lilac-coloured flowers immediately distinguish Field Madder from other cleavers.

Associated uncommon species: Field Madder is frequently found in species-rich arable plant communities on chalk or limestone soils with species including Rough Poppy, Night-flowering Catchfly, Field Gromwell and Pheasant's-eye.

HABITAT:
Arable field margins, horticultural land, tracks, disturbed grassland.

SOIL TYPE:
Light calcareous loams.

DISTRIBUTION:
Field Madder is widespread throughout lowland Britain as far north as eastern Scotland.

LIFE CYCLE:
Flowers from May to September, often appearing in stubbles after harvest.
Seed longevity is little known. Germination is in both autumn and spring.

J F M A M J J A S O N D

REASONS FOR DECLINE:
Field Madder is probably susceptible to many herbicides, and is a non-competitive species that does not grow well in a fully-fertilised modern crop variety. Because it also flowers in stubbles, is has probably been affected by early ploughing.

Field Madder: fruit ×6

Corn Marigold

Chrysanthemum segetum

IDENTIFICATION:

Corn Marigold is usually a robust, much-branched plant growing up to 1 m in height on nitrogen-rich soils. The leaves are around 10 cm long, deeply to shallowly lobed. The plant is hairless with a waxy surface, giving it a blue-green appearance. The flower-heads are 3–6 cm across and, typical of the daisy family (Asteraceae), resemble a single, large flower. The central disc-florets are golden yellow, whilst the ray-florets making up the outer part of the flower-head have single golden petals up to 1·5 cm long. The seeds are of two types: those of the inner disc-florets are cylindrical and un-winged, whilst those of the outer ray-florets are broad and winged.

Similar species: The golden-yellow flowers and the blue-green colour of the whole plant make Corn Marigold unmistakable in arable habitats.

Associated uncommon species: Corn Marigold can sometimes grow in species-rich communities in south-west England and Wales, with Small-flowered Catchfly, Lesser Quaking-grass, Weasel's-snout and, rarely, Rough Poppy and Night-flowering Catchfly.

HABITAT:
Arable fields.

SOIL TYPE:
Sands and sandy loams, rarely on other soils in the west of Britain. These soils are sometimes quite calcareous when derived from shell-sand.

MANAGEMENT REQUIREMENTS:
Spring cultivation.

DISTRIBUTION:
Corn Marigold is widely distributed but has declined considerably within its range since the 1970s. Now most frequent in the west of Britain, as far north as Shetland, but still occurs on isolated sites elsewhere.

LIFE CYCLE:
Flowers from June to August.
Seed can be very long-lived.
Germination can take place in the autumn, but plants are usually killed by frost unless they establish well before the winter.

J F M A M J J A S O N D

REASONS FOR DECLINE:
The development of effective herbicides in the 1970s is a significant factor in the decline of Corn Marigold as well as, in western Britain, the conversion of arable land to pasture and the reduction in the area of spring cereals grown.

Corn Marigold: leaf × 2

Mousetail

Myosurus minimus

IDENTIFICATION:

This member of the buttercup family (Ranunculaceae) is very distinctive. The plant consists of a rosette of very narrow, linear, dark green leaves up to 85 mm long. There are from one to many leafless flowering stems growing from the centre of the rosette. These flowering stems grow to 15 cm in height, rarely taller, and are sometimes as short as 2 cm. The flowers are initially inconspicuous with small, green-yellow petals which soon fall. The seed-bearing receptacle elongates after flowering to form a long, narrowly conical and conspicuous 'mouse's tail'.

Similar species: This species is unmistakable.

Associated uncommon species: Mousetail has occurred at some sites with Corn Buttercup.

HABITAT

Mousetail often grows in field gateways, muddy trackways and hollows in arable fields. It also occurs in disturbed and trampled grassland and woodland rides.

SOIL TYPE:

Mousetail grows in areas where water accumulates during the winter, either on compacted clay soils or on gravelly soils overlying clays. These soils are frequently non-calcareous.

MANAGEMENT REQUIREMENTS:

Autumn cultivation.

DISTRIBUTION:

Occurs widely in central-southern and eastern England. There are centres of distribution in Essex and the south-west Midlands.

LIFE CYCLE:

Flowers from May to June.
Seed is long-lived.
Germination is entirely in the autumn and early winter.

J F M A M J J A S O N D

REASONS FOR DECLINE:

This species has probably declined as a result of field drainage, the surfacing of farm tracks and the use of more competitive crop varieties and higher nitrogen application levels. The lack of grazing of commons may also have led to the loss of some sites.

Mousetail: flower and seed spike ×2

Field Pansy

Viola arvensis

IDENTIFICATION:

Field Pansy resembles a diminutive version of a garden pansy. It is an erect, branched plant, sometimes scrambling among cereal plants to a height of 40 cm, but often much smaller. Individual plants can become quite large. The leaves are highly variable in shape. The lower leaves are oval, becoming narrower towards the top of the plant. The upper leaves are narrowly oval or spear-shaped. There are also large leafy stipules at the base of the true leaves, deeply divided into narrow lobes. The bilaterally symmetrical flowers have a short spur, measure 8–20 mm from top to bottom and are borne singly on long stems. The petals are separate and inclined towards each other to form a shallow cup. The petals are generally cream in colour with a yellow centre, the upper petals are sometimes pale blue or purple. The sepals are as long as the petals, and persist after flowering, enclosing the seed capsule.

Similar species: The only other similar species is the Wild Pansy from which it can be difficult to distinguish, particularly as the two species sometimes hybridise. The flowers of Wild Pansy are larger, flat rather than cupped and with upper petals that are frequently blue or violet. The sepals are shorter than the petals.

Associated uncommon species: Field Pansy is frequently part of arable plant communities including uncommon species.

HABITAT:

Arable fields.

SOIL TYPE:

Well-drained soils.

DISTRIBUTION:

Field Pansy is present throughout Britain, it is mainly a lowland species and is less common in the north.

LIFE CYCLE:

Flowers from June until fields are ploughed or until the first frosts. Seed can persist in the soil for many years.
Germination occurs both in spring and autumn.

J F M A M J J A S O N D

Field Pansy: flower × 1

REASONS FOR DECLINE:

Field Pansy has remained common in recent years.

Wild Pansy

Viola tricolor

IDENTIFICATION:

Wild Pansy resembles a diminutive version of a garden pansy. It is an erect, branched plant, sometimes reaching a height of 40 cm, but usually much smaller. The leaves are highly variable in shape. The lower leaves are oval, becoming narrower towards the top of the plant. The upper leaves are narrowly oval or spear-shaped. There are also large leafy stipules at the base of the true leaves, deeply divided into narrow lobes. The bilaterally symmetrical flowers have a short spur, measure 15–25 mm from top to bottom and are borne singly on long stems. The face of the flower is flat. The lower petals are generally cream in colour with a yellow centre; the upper petals are pale blue or purple. The sepals are shorter than the petals, and persist after flowering, enclosing the seed capsule.

Similar species: The only other similar species is the Field Pansy from which it can be difficult to distinguish, particularly as the two species sometimes hybridise. Field Pansy flowers are smaller, cupped rather than flat, and the upper petals are usually cream (although sometimes blue or violet) rather than the pale blue or purple of Wild Pansy. In addition, Field Pansy has sepals as long as the petals.

Associated uncommon species: Wild Pansy is frequently part of arable plant communities including other uncommon species.

HABITAT:

Arable field margins, sandy grassland, dunes.

SOIL TYPE:

Sandy soils.

DISTRIBUTION:

Wild Pansy is present throughout Britain, but less common in the south.

LIFE CYCLE:

Flowers from June to September. Seed can persist in the soil for many years.
Germination occurs both in spring and autumn.

J F M A M J J A S O N D

REASONS FOR DECLINE:

Wild Pansy is a much less competitive plant than Field Pansy, and has probably been disadvantaged by increased applications of nitrogen to modern crop varieties.

Wild Pansy: flower × 1

Corn Parsley

Petroselinum segetum

IDENTIFICATION:

Corn Parsley is a rather unusual member of the carrot family (Apiaceae). It can grow up to 80 cm tall with a much-branched stem. The whole plant is dark green and hairless, and smells of parsley if crushed. The leaves are narrowly spear-shaped and are pinnately divided into lobed segments. The small, white-petalled flowers are borne in branched clusters typical of its family. There can be 3–5 rays of highly unequal length (2–40 mm) in each cluster, and overall the flowering parts present an extremely irregular appearance.

Similar species: The only similar species is the closely-related Stone Parsley, which grows in damp roadsides, occasionally disturbed wet areas and ditches, sometimes bordering arable fields. Stone Parsley differs in having an unpleasant 'petrol' scent.

Associated uncommon species: In arable land it is often part of species-rich communities on chalky soils, including Pheasant's-eye, Rough Poppy, Field Gromwell and Narrow-fruited Cornsalad, and on limestones, including Broad-leaved Spurge, Corn Buttercup and Spreading Hedge-parsley.

HABITAT:

Corn Parsley is found in arable field margins and open hedge bottom vegetation. It is also found in parched grasslands on clay banks near the sea, on sea-walls, Cornish hedges and on road verges.

SOIL TYPE:

Calcareous clays and loams. Particularly frequent on clay soils over Jurassic limestones in south-west England.

MANAGEMENT REQUIREMENTS:

This species does best where autumn-sown crops are followed by spring-sown crops with a stubble left through the autumn.

Corn Parsley: fruit × 5

DISTRIBUTION:

Scattered throughout southern Britain northwards to Lincolnshire. Most frequent around the coast, but there are also concentrations in mid-Somerset, the Wessex chalk and the North Downs.

LIFE CYCLE:

Flowers from August to September. Seed longevity unknown.
Germination is almost entirely in September and October.

J F M A M J J A S O N D

REASONS FOR DECLINE:

Corn Parsley has a long growing season, germinating in early autumn and flowering in August and September. It has probably been affected by earlier harvesting dates and the rarity of autumn stubbles. It is probably susceptible to many herbicides.

Field Penny-cress
Thlaspi arvense

IDENTIFICATION:
Field Penny-cress is an erect, hairless, sparingly-branched plant, up to 60 cm in height. The stem-leaves are oblong in outline, up to 7 cm long, with backward-pointing auricles that clasp the stem. The flowers are symmetrical, 4–6 mm in diameter with four white petals. Many flowers are borne in an inflorescence at the top of the stem, which becomes much longer after flowering. The conspicuous and characteristic fruits are on upwardly-curving stalks 5–15 mm long. The fruits are circular, 12–20 mm across and consist of a central part containing the seeds which is encircled by broad wings.

Similar species: The characteristic shape of the fruit distinguishes this from all similar species. The fruit of Perfoliate Penny-cress is heart-shaped and only 4–6 mm long.

Associated uncommon species: Field Penny-cress is rarely associated with uncommon species.

HABITAT:
Field Penny-cress is found largely in arable field margins and occasionally in gardens, roadsides and other disturbed sites.

SOIL TYPE:
Most frequent on neutral loamy and clay soils.

DISTRIBUTION:
Present throughout the lowlands of Britain. It is most common in the east.

LIFE CYCLE:
Flowers from May to July.
Seed is long-lived in the soil.
Germination is mostly in the spring between March and May.

J F M A M J J A S O N D

REASONS FOR DECLINE:
None. This species remains frequent over much of its range, and has increased in some areas.

Field Penny-cress: seed × 2

Perfoliate Penny-cress
Thlaspi perfoliatum

IDENTIFICATION:
Perfoliate Penny-cress has erect, leafy stems up to 30 cm tall. The whole plant is hairless and bluish-green in colour. The stem-leaves are oval, with basal lobes clasping the stem. The flowers are borne in short, terminal spikes. The flowers have narrow, white petals and are 2·0–2·5 mm in diameter. The fruits are flattened and heart-shaped, held parallel to the ground. They are approximately 5 mm long by 4 mm wide with membranous wings along each edge. The seeds are yellowish-orange.

Similar species: The only similar species is the common Field Penny-cress. This has much larger, circular fruits 12–20 mm across, and with light-green leaves that only partly clasp the stem.

Associated uncommon species: None known at its arable site.

HABITAT:
Drought-prone, disturbed grasslands, in arable field margins at only one known site.

SOIL TYPE:
Calcareous clay loams over Jurassic limestones.

MANAGEMENT REQUIREMENTS:
It is likely to grow best after autumn cultivation.

DISTRIBUTION:
Confined to Oxfordshire, Gloucestershire and Worcestershire, although there are casual occurrences elsewhere.

LIFE CYCLE:
Flowers from April to May. Seed is relatively short-lived. Germination is mainly in the early autumn, although some seedlings can also appear in the spring.

J F M A M J J A S O N D

REASONS FOR DECLINE:
Perfoliate Penny-cress has always been very rare in arable habitats in Britain.

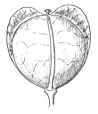

Perfoliate Penny-cress: seed ×5

Pheasant's-eye

Adonis annua

IDENTIFICATION:

Pheasant's-eye is a member of the buttercup family (Ranunculaceae). Plants can grow up to 50 cm in height and can be much-branched. The leaves are finely dissected with bright-green, narrow, parallel-sided segments. The flower resembles that of an anemone. Petals are deep red with a dark basal spot and the anthers are black. The flower is 15–25 mm across. There are rarely more than 30 flowers per plant, and the elongated oval seed head usually bears about 30 seeds. The seeds are similar in size and shape to a grape-pip, and are olive-green in colour.

Similar species: Unmistakable when flowering; young plants can be mistaken for mayweeds but Pheasant's-eye differs in being a much brighter green colour, completely hairless and with broader leaf segments.

Associated uncommon species: Always in species-rich communities with Corn Parsley and Rough Poppy, often with Narrow-fruited Cornsalad, Spreading Hedge-parsley, Prickly Poppy, Dense-flowered and Small-flowered Fumitories, Night-flowering Catchfly and Red Hemp-nettle.

HABITAT:

Arable land. One population is known from rides in a forestry plantation created on arable land in the 1940s, and it is also found in disturbed ground on former arable land on Salisbury Plain.

SOIL TYPE:

Always calcareous, either on chalk or on oolitic limestone in Dorset and Gloucestershire. Soil texture ranges from silty loams to clay loams.

Pheasant's-eye: seed head (*left*) × 1, seed (*right*) × 3

DISTRIBUTION:

As a native plant, Pheasant's-eye has always been confined to the chalk and limestones of southern England. It has never been widespread although it was sufficiently abundant in Sussex in the mid-18th century to have been collected for sale as a cut flower.

LIFE CYCLE:

Flowers from June to July.
Seed is thought to be quite long-lived.
Germination is mainly in the autumn, with some in the spring.

J F M A M J J A S O N D

REASONS FOR DECLINE:

Pheasant's-eye is highly sensitive to herbicides. It is also poorly competitive in relation to modern crops, and high levels of nitrogen use will have affected it. Few seeds are produced, and seed-banks are probably quite small.

Greater Pignut

Bunium bulbocastanum

IDENTIFICATION:

A typical member of the carrot family (Apiaceae) growing up to 60 cm tall. It is a perennial, over-wintering as an underground tuber that forms at the base of the stem. The leaves are very finely, 3-pinnately divided into narrow, 5–10 mm wide, segments. The leaves are long-stalked with a base that sheathes the stem. The flowers are white, and are borne in a dense, umbrella-shaped cluster, 30–80 mm across, with several small, undivided bracts at the base. The brown fruits are oval and approximately 3 mm long.

Similar species: Many other species in the same family are superficially similar, but Greater Pignut can be distinguished by its characteristically divided leaves. The similar Common Pignut, differs in having a hollow stem, is usually found in woods and is rarely found on the soil type preferred by Greater Pignut.

HABITAT:

Irregularly-cultivated arable field edges, tracksides, disturbed chalk grasslands.

SOIL TYPE:

Calcareous loams over chalk.

MANAGEMENT REQUIREMENTS:

Occasional disturbance of field edge habitats, and removal of any invading scrub.

DISTRIBUTION:

North Chilterns from Buckinghamshire to Cambridgeshire.

Greater Pignut: fruit × 10

LIFE CYCLE:

Flowers from June to July.
Seed longevity is unknown.
Germination period unknown.

J	F	M	A	M	J	J	A	S	O	N	D
?										?	

REASONS FOR DECLINE:

Intensive cultivation of field edge habitats.

170

Blue Pimpernel

Anagallis arvensis ssp. *foemina*

IDENTIFICATION:

The common Scarlet Pimpernel (INSET) usually has scarlet flowers, but does have a blue-flowered form (LOWER). However, Blue Pimpernel (UPPER) is a distinct subspecies rather than a colour form. It is a much-branched, scrambling or mat-forming plant with stems up to 30 cm in length. The leaves measure up to 20 mm × 9 mm and are stalkless, narrowly oval in shape, and pointed at the apex with tiny black glands underneath. They are arranged in opposite pairs along the stem. The blue, 5-petalled flowers are approximately 10 mm in diameter and are borne singly in leaf axils on stalks that are shorter than the leaves. The petals do not overlap one another and the sepals are longer than the petals. When in bud, the sepals conceal the petals completely. The seeds are produced in spherical capsules 4 mm in diameter on recurved stalks.

Similar species: The blue-flowered form of Scarlet Pimpernel is very similar and has been much confused in the past. This form has larger, broader petals that overlap, a flower stalk that is longer than the associated leaf and sepals that are shorter than the petals and which, when in bud, do not completely conceal the petals.

Associated uncommon species: Blue Pimpernel is normally found in species-rich arable plant communities including Night-flowering Catchfly, Field Gromwell, Spreading Hedge-parsley and Slender Tare.

HABITAT:

Arable field margins on south-facing slopes.

SOIL TYPE:

Blue Pimpernel is found mainly on well-drained, calcareous soils. Most sites are on Jurassic limestones.

DISTRIBUTION:

Largely confined to the south of England.

LIFE CYCLE:

Flowers from June to August.
The closely related Scarlet Pimpernel has long-lived seeds which germinate almost entirely in the spring in April and May.

J F M A M J J A S O N D

REASONS FOR DECLINE:

Distribution of Blue Pimpernel has been poorly known in the past. There is likely to have been a decline due to a combination of the high levels of nitrogen applied to competitive modern crops, the use of broad-spectrum herbicides and increases in the proportion of the cropped area sown with winter cereals.

Blue Pimpernel: seed capsule × 2

Babington's Poppy

Papaver dubium ssp. *lecoqii*

IDENTIFICATION:
Babington's Poppy rarely grows taller than 60 cm. It has a basal rosette of leaves and several stem-leaves. The basal leaves die as the plant matures. Leaves are shallowly lobed or pinnately divided into broad, spear-shaped segments, and are blue-green in colour. The plants can branch from the base but more usually it is the main stem that is branched. The flowers are borne at the end of long stems covered with tightly-pressed hairs. The flowers of Babington's Poppy are usually up to 5 cm in diameter with four pinkish-red, unblotched petals, which usually only persist for a single day. The anthers are yellow to brown in colour. The seed capsule is elongated, without prickles. If the stem or leaves are broken, the exuded latex turns yellow on contact with the air.

Similar species: Other poppies, especially Long-headed Poppy. Only Babington's and Long-headed Poppies have an elongated bristle-free seed capsule. Long-headed Poppy can be distinguished by its different coloured anthers and by its latex which remains white on contact with the air.

HABITAT:
Arable field margins and road verges.

SOIL TYPE:
Heavy, calcareous loams, rarely on chalk.

DISTRIBUTION:
Babington's Poppy is found mainly in the south and east of England, but its distribution is poorly known.

Babington's Poppy: seed capsule ×2

LIFE CYCLE:
Flowers from June to July.
Seed is very long-lived.
Germination is from October to November and from March to April.

J F M A M J J A S O N D

REASONS FOR DECLINE:
Increases in levels of nitrogen application and the use of herbicides.

Common Poppy

Papaver rhoeas

IDENTIFICATION:

Common Poppy grows up to 80 cm. It has a basal rosette of leaves and several stem-leaves. The basal leaves die as the plant matures. The leaves are blue-green in colour and shallowly lobed or pinnately divided into broad, spear-shaped segments. The plants branch both from the base and the main stem. The flowers are borne at the end of long stems covered with erect hairs. Flowers of Common Poppy have four bright scarlet petals, frequently blotched with black at the base and can measure up to 8 cm in diameter. Petal colour can be highly variable, from white to deep red, and even variegated. The flowers can also be compound. This variability is the origin of the garden poppy varieties. The anthers are blue-black in colour. Large plants can have several hundred flowers; small plants can have as few as one. The seed capsule is essentially spherical with a flattened top, the seeds being shed through openings in a disc covering the top of the capsule. The capsule has no prickles.

Similar species: Other poppies. Common Poppy can be distinguished from all other species by the spherical, bristle-free seed capsule.

Associated uncommon species: Common Poppy occurs in a wide range of communities, and is often present where rare species occur.

HABITAT:

Arable fields, disturbed ground and waste places.

SOIL TYPE:

It is found on a wide range of soil types, but it is most common on well-drained calcareous loams.

Common Poppy: seed capsule ×2

DISTRIBUTION:

Common Poppy is found throughout Britain, but it is most frequent in the south and east of England.

LIFE CYCLE:

Flowers from June to August. Seeds are very long-lived. Germination is in both autumn and spring.

J F M A M J J A S O N D

REASONS FOR DECLINE:

Common Poppy is susceptible to many herbicides, including many of the earliest compounds. Increases in levels of nitrogen application have also been important. The seed longevity means that large numbers of plants can appear following reduced intensification of farming practices, even where no plants have been seen for many years.

Long-headed Poppy
Papaver dubium

IDENTIFICATION:
Long-headed Poppy rarely grows taller than 60cm. It has a basal rosette of leaves and several stem-leaves, the basal leaves dying as the plant matures. Leaves are blue-green in colour and shallowly lobed or pinnately divided into broad, spear-shaped segments. The plants can branch from the base but more usually it is the main stem that is branched. The flowers are borne at the end of long stems covered with tightly-pressed hairs. Flowers of Long-headed Poppy are usually up to 5 cm in diameter with four pinkish-red, unblotched petals, which usually only persist for a single day. The anthers are brown to blue-black in colour. The seed capsule is elongated up to 2·5 cm, without prickles. If the stem or leaves are broken, the exuded latex remains white.

Similar species: Other poppies, especially Babington's Poppy. Only Long-headed and Babington's Poppies have an elongated bristle-free seed capsule. Babington's Poppy can be distinguished by its different coloured anthers and by its latex which turns yellow on contact with the air.

Associated uncommon species: Long-headed Poppy occurs in a wide range of communities, and is often present where rare species occur.

HABITAT:
Wide range.

SOIL TYPE:
It is found on a wide range of soil types.

DISTRIBUTION:
Long-headed Poppy is found throughout Britain, but it is most frequent in the south and east of England.

Long-headed Poppy: seed capsule ×2

LIFE CYCLE:
Flowers from May to August. Seeds are very long-lived. Germination is largely between September and November but it can also germinate in the spring.

J F M A M J J A S O N D

REASONS FOR DECLINE:
Increases in levels of nitrogen application and the use of herbicides.

Prickly Poppy

Papaver argemone

IDENTIFICATION:

Prickly Poppy rarely grows taller than 50 cm. It has a basal rosette of leaves and several stem-leaves. The basal leaves die as the plant matures. Leaves are pinnately divided into narrow, parallel-sided segments. The plants can branch from the base but more frequently it is the main stem that is branched. The flowers are borne at the end of long stems that are covered with short, tightly-pressed hairs. Flowers of Prickly Poppy are usually up to 5 cm in diameter with four distinctive orange-red petals which usually only persist for a single day. The seed capsule is elongated and club-shaped, usually 2·0–2·5 cm long with numerous prickles and with a terminal disc. The stamens surround the capsule, and the anthers have blue pollen.

Similar species: Other poppies. Prickly Poppy can be distinguished from species other than Rough Poppy by the narrow, parallel-sided leaf segments. When flowering, the orange-red petals are very distinctive. The long, narrow, prickly seed capsule distinguishes Prickly Poppy from all other species.

Associated uncommon species: Especially on chalky soils, Prickly Poppy is often a member of species-rich communities with Rough Poppy, Field Gromwell, Dense-flowered Fumitory, Narrow-leaved Cornsalad and Shepherd's-needle.

SOIL TYPE:

Most frequently on free-draining chalky and sandy loams. Also on calcareous clay loams.

Prickly Poppy: seed capsule ×2

DISTRIBUTION:

Formerly a frequent species throughout England, particularly in the south and east, extending northwards to isolated centres of population in eastern Scotland. It has gone from many of its northern and western sites, and has become increasingly restricted to chalk and sands in south-east England.

LIFE CYCLE:

Flowers from June to July.
Seed is very long-lived.
Germination is largely from September to November and from March to April.

J F M A M J J A S O N D

REASONS FOR DECLINE:

Increases in levels of nitrogen application and the use of herbicides.

Rough Poppy

Papaver hybridum

IDENTIFICATION:

Like Prickly Poppy, Rough Poppy rarely grows taller than 50 cm. It has a basal rosette of leaves and several stem-leaves. The basal leaves die as the plant matures. Leaves are pinnately divided into narrow, parallel-sided segments. The plants can branch from the base but more frequently it is the main stem that is branched. The flowers are borne at the end of long stems that are covered with spreading hairs. Flowers of Rough Poppy usually only persist for a single day and can be up to 5 cm across, with four distinctively scarlet petals that have black blotches at their base. The seed capsule is 1·0–1·5 cm in diameter, spherical with numerous upward-pointing prickles and a terminal disc. The stamens surround the capsule, and the anthers have deep-blue pollen.

Similar species: Other poppies. Rough Poppy can be distinguished from species other than Prickly Poppy by the narrow, parallel-sided leaf segments. When flowering, the scarlet petals are very distinctive. The spherical, bristly seed capsule distinguishes Rough Poppy from all other species.

Associated uncommon species: Especially on chalky soils, Rough Poppy is often a member of species-rich communities with Prickly Poppy, Pheasant's-eye, Field Gromwell, Dense-flowered Fumitory, Narrow-leaved Cornsalad and Shepherd's-needle.

SOIL TYPE:

Most frequently on chalky loams, sometimes on other calcareous soils including boulder clay.

DISTRIBUTION:

Rough Poppy is found mainly in the south and east of England, particularly on the chalk outcrops of central-southern England, the North and South Downs, the Chilterns and Cambridgeshire. There are outlying sites in north Cornwall and Norfolk, with recent records as far north as Lincolnshire.

LIFE CYCLE:

Flowers from June to July.
Seed is very long-lived.
Germination is from October to December and from March to April.

J F M A M J J A S O N D

REASONS FOR DECLINE:

Increases in levels of nitrogen application and the use of herbicides.

Rough Poppy: seed capsule ×2

Shepherd's-needle

Scandix pecten-veneris

IDENTIFICATION:

When flowering, Shepherd's-needle is a typical member of the carrot family (Apiaceae). It grows up to 60 cm tall when supported by the crop, with an occasionally-branched stem. The plant has very few hairs and is a bright green colour before the fruit begins to ripen. The leaves are finely divided, with narrow, parallel-sided segments. The small, white-petalled flowers are in umbrella-like clusters, with several bisected bracts at the base. The fruit attains its characteristic form shortly after flowering begins. Each flower forms two seeds joined together. Each seed has a long needle-like projection up to 5 cm long that acts as a spring as the fruits mature, flicking them up to 1m from the parent plant. The 'needle' also has minute bristles that can attach the seed to fur or clothing and help distribution.

Similar species: Shepherd's-needle can be readily distinguished from other related species by the unique fruit, and, before fruiting, by the bisected bracts.

Associated uncommon species: Often in species-poor communities on heavy soils. It can sometimes occur with uncommon species including Spreading Hedge-parsley, Corn Buttercup, Broad-leaved Spurge, Corn Parsley and Field Gromwell.

HABITAT:

Arable fields, occasionally on dry sunny banks near the coast.

SOIL TYPE:

Most frequent on heavy calcareous clay loams, but sometimes on a wide range of other soil types.

MANAGEMENT REQUIREMENTS:

Autumn cultivation.

Shepherd's-needle: seed, seed head × 1

DISTRIBUTION:

Once abundant in eastern Britain, north to the Moray Firth, by the mid-1980s it was very rare, but since 2000 has recovered, appearing at new sites. Although abundant in parts of East Anglia, it is still very rare in the west.

LIFE CYCLE:

Flowers from April to July.

Seed is very short-lived.

Germination is largely from October to December, but a few seedlings also appear in spring.

J F M A M J J A S O N D

REASONS FOR DECLINE:

Susceptible to many broad-spectrum herbicides. Its recent reappearance may be due to the over-use of the few herbicides to which this plant is resistant. Where winter crops are grown continuously, the end of stubble burning may also have helped this species, as it may now be spread between farms in straw.

Breckland Speedwell

Veronica praecox

IDENTIFICATION:

Breckland Speedwell has an erect, sometimes branched, flowering stem growing up to 20 cm tall although frequently much smaller. The leaves are toothed but not deeply divided. The dark-streaked, blue flowers are small, 3 mm in diameter, with long stalks, and are borne in the axils of leaf-like bracts. The fruit capsule is bi-lobed, longer than wide.

Similar species: Ivy-leaved Speedwell and Field-speedwells including Common, Green, and Grey form trailing mats and have creeping, rather than erect, stems. Fingered, Spring and Wall Speedwells are other species with erect stems and differ as follows:

FINGERED SPEEDWELL – has leaves divided into three parallel lobes.

SPRING SPEEDWELL – has pinnately-lobed leaves with 5–7 segments.

WALL SPEEDWELL – has toothed leaves but the flowers are without stalks.

Associated uncommon species: Other Breckland rarities including Fingered Speedwell.

HABITAT:

Arable field margins; disturbed sandy grassland.

SOIL TYPE:

Sands, frequently calcareous.

MANAGEMENT REQUIREMENTS:

Occasional autumn cultivation of field edges and tracksides.

DISTRIBUTION:

Confined to a few sites in Breckland. Also at one site in Oxfordshire.

LIFE CYCLE:

Flowers from March to May. Seed is probably quite long-lived. Germination is in the autumn.

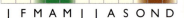

J F M A M J J A S O N D

REASONS FOR DECLINE:

Intensive cultivation of field edges, increased application of nitrogen and the use of much more competitive crop varieties.

Breckland Speedwell: leaf × 2

Fingered Speedwell

Veronica triphyllos

IDENTIFICATION:

Fingered Speedwell is a low-growing, occasionally-branched plant, rarely reaching as tall as 15 cm. The leaves are up to 1 cm long, deeply divided into 3–7 parallel-fingered lobes. The upper leaves are stalkless, whereas the lower leaves have short stalks. The flowers are borne in the axils of the upper leaf-like bracts. They are 3–4 mm in diameter, deep blue and shorter than the surrounding calyx. The fruit capsule is deeply bi-lobed, about 6–7 mm long.

Similar species: Ivy-leaved Speedwell and Field-speedwells including Common, Green, and Grey form trailing mats and have creeping, rather than erect, stems. Breckland, Spring and Wall Speedwells are other species with erect-stems and differ as follows:

BRECKLAND SPEEDWELL – has toothed leaves.

SPRING SPEEDWELL – has pinnately-lobed leaves with 5–7 segments, and very small flowers.

WALL SPEEDWELL – has toothed leaves.

Associated uncommon species: Other typical Breckland rarities.

HABITAT:

Arable field margins, fallows and disturbed parched grassland.

SOIL TYPE:

Sands.

MANAGEMENT REQUIREMENTS:

Occasional autumn cultivation of field edges and tracksides.

DISTRIBUTION:

Fingered Speedwell has always been a rare species restricted to a few sites in East Anglia and Yorkshire. It is now known as a native species from only two sites in the Breckland of Norfolk and Suffolk. It is also present as an introduction at another Suffolk site.

LIFE CYCLE:

Flowers from March to May.
Seed longevity is unknown.
Germination is in the early autumn.

J F M A M J J A S O N D

REASONS FOR DECLINE:

Fingered Speedwell is a poorly competitive species that has been badly affected by the high levels of nitrogen applied to modern crop varieities. Other speedwells are very susceptible to a wide variety of herbicides. Several sites have been lost to development.

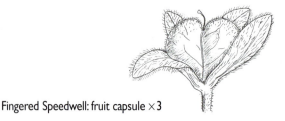

Fingered Speedwell: fruit capsule × 3

Spring Speedwell

Veronica verna

IDENTIFICATION:

Spring Speedwell has an erect, sometimes branched flowering stem growing up to 15 cm tall although frequently much smaller. The leaves are pinnately-lobed with 5–7 segments. The sky-blue flowers are very small, only 2–3 mm in diameter, with short stalks, and are borne in the axils of leaf-like bracts. The fruit capsule is bi-lobed, wider than long.

Similar species: Ivy-leaved Speedwell and Field-speedwells including Common, Green, and Grey form trailing mats and have creeping, rather than erect, stems. Breckland, Fingered and Wall Speedwells are other species with erect stems and differ as follows:

BRECKLAND SPEEDWELL – has toothed leaves.

FINGERED SPEEDWELL – has leaves divided into three parallel lobes, and larger flowers.

WALL SPEEDWELL – has toothed leaves.

Associated uncommon species: Other Breckland rarities including Fingered Speedwell.

HABITAT:

Arable field margins, disturbed sandy grassland and tracksides.

SOIL TYPE:

Sands, frequently calcareous.

MANAGEMENT REQUIREMENTS:

Occasional autumn cultivation of field edges and tracksides.

DISTRIBUTION:

Spring Speedwell is restricted to a few sites in the Suffolk Breckland.

LIFE CYCLE:

Flowers from May to September. Seed is probably quite long-lived. Germination is in the autumn.

J F M A M J J A S O N D

REASONS FOR DECLINE:

Intensive cultivation of field-edges, increased application of nitrogen and the use of much more competitive crop varieties.

Spring Speedwell: fruit capsule (*left*) × 3, flower (*right*) × 10

Green Field-speedwell

Veronica agrestis

IDENTIFICATION:

Green Field-speedwell is a sprawling plant with stems that are softly hairy and much branched. It can sometimes form extensive mats in autumn stubble. The leaves are green, oval in shape with toothed edges, 5–15 mm across and slightly longer than wide. The solitary flowers are borne on stalks in the leaf axils. They are usually approximately 5 mm in diameter and pale blue with a white centre and lower lobe. The fruit capsule is bi-lobed, 3–4 mm × 4–6 mm, wider than long and with erect, rather than spreading, lobes.

Similar species: Common Field-speedwell and Grey Field-speedwell are very similar, with creeping stems that form trailing mats. Common Field-speedwell has larger, 1–3 cm, leaves; larger, 8–12 mm, bright blue flowers on longer stalks; and larger, 4–6 mm × 7–8 mm, spreading fruit lobes. Grey Field-speedwell has intense blue flowers and greyish-green leaves.

Associated uncommon species: Green Field-speedwell can sometimes be found with species such as Corn Marigold and Weasel's-snout.

HABITAT:

Arable field margins.

SOIL TYPE:

Well-drained sandy loam soils, usually non-calcareous.

DISTRIBUTION:

Green Field-speedwell is present throughout Britain to the Shetland Islands, but is most common in the western half of the country.

LIFE CYCLE:

Flowers from June until the stubble is ploughed or until killed by frost.
Seed biology is little known.
Germination appears to be mainly in the spring.

J F M A M J J A S O N D

REASONS FOR DECLINE:

Green Field-speedwell is probably susceptible to many herbicides, and is a non-competitive species that does not grow well in a fully-fertilised modern crop variety. Because it also flowers in stubbles, it has probably been affected by early ploughing.

Green Field-speedwell: fruit capsule ×4

Grey Field-speedwell
Veronica polita

IDENTIFICATION:
Grey Field-speedwell is a sprawling plant with stems that are much-branched. It can sometimes form extensive mats in autumn stubble. The leaves are a dull greyish-green colour, oval in shape with toothed edges and 5–15 mm wide. The solitary, asymmetrical flowers are borne on stalks in the leaf axils. They are usually about 5 mm in diameter, bright blue with a white centre, and with the lower lobe occasionally paler. The fruit capsule is bi-lobed; 3–4 mm × 4–6 mm, wider than long and with erect, rather than spreading, lobes.

Similar species: Common Field-speedwell and Green Field-speedwell are very similar, with creeping stems that form trailing mats. Common Field-speedwell has larger, 1–3 cm, leaves; larger, 8–12 mm, bright blue flowers on longer stalks; and larger, 4–6 mm × 7–8 mm, spreading fruit lobes. Green Field-speedwell has pale blue flowers and green leaves.

Associated uncommon species: Grey Field-speedwell can often be found in species-rich arable plant communities with uncommon species including Rough Poppy, Field Gromwell, Red Hemp-nettle and Pheasant's-eye.

HABITAT:
Arable field margins.

SOIL TYPE:
Well-drained loamy soils, frequently calcareous.

DISTRIBUTION:
Grey Field-speedwell is present throughout lowland Britain northwards to southern Scotland, but is most common in the south and east of England.

LIFE CYCLE:
Flowers from June until the stubble is ploughed or until killed by frost. Seed is probably quite long-lived. Germination appears to be mainly in the spring.

J F M A M J J A S O N D

REASONS FOR DECLINE:
Grey Field-speedwell is probably susceptible to many herbicides, and is a non-competitive species that does not grow well in a fully-fertilised modern crop variety. Because it also flowers in stubbles it has probably been affected by early ploughing.

Grey Field-speedwell: fruit capsule × 4

Broad-leaved Spurge

Euphorbia platyphyllos

IDENTIFICATION:

Broad-leaved Spurge can grow up to 60 cm tall. Leaves are arranged alternately along the stem. They are narrowly oblong, up to 5 cm long, pointed at the tip, and with serrated edges. The leaves are hairless or lightly hairy, and the stems are often reddish. The unusual, yellowish-green flowers, like all spurges, lack petals, and are arranged in a loose umbel with 3–5 rays at the stem apex. The base of the umbel is ringed by leafy bracts. Each flower produces a roughly spherical fruit capsule approximately 3 mm in diameter and covered in small warts. This fruit splits explosively on ripening.

Similar species: The only similar species are Sun Spurge and Petty Spurge but neither of these species have serrated leaves or warty fruits. In addition, Sun Spurge lacks stem-leaves.

Associated uncommon species: Broad-leaved Spurge is usually found in species-rich communities with species including Corn Buttercup, Shepherd's-needle, Broad-fruited and Narrow-fruited Cornsalads and Corn Parsley.

HABITAT:

Arable field margins.

SOIL TYPE:

Calcareous clay loams or silty loams.

MANAGEMENT REQUIREMENTS:

Autumn cultivation.

Broad-leaved Spurge: fruit capsule ×8

DISTRIBUTION:

Formerly present as far north as Yorkshire, Broad-leaved Spurge has become much less frequent in recent years. It now occurs across the south of England from Somerset to Cambridgeshire, but is most frequent in Somerset.

LIFE CYCLE:

Flowers from June to October.
Seed is thought to be long-lived in the soil, but relatively few seeds are produced.
Germination is mainly in the autumn, but seedlings can also germinate in the spring.

J F M A M J J A S O N D

REASONS FOR DECLINE:

Broad-leaved Spurge is susceptible to many herbicides and it also poorly competitive with highly fertilised modern crop varieties.

Dwarf Spurge

Euphorbia exigua

IDENTIFICATION:

Dwarf Spurge is a light green, hairless plant, usually no more than 10 cm tall, and sometimes much smaller. The narrow, parallel-sided leaves are up to 2 cm long and 3 mm wide, rarely larger, and are arranged alternately along the stem. The unusual, yellowish-green flowers lack petals, and are arranged in a loose umbel with 3–5 rays at the stem apex. The base of the umbel is ringed by leafy bracts. Each flower produces a nearly spherical fruit capsule approximately 2 mm in diameter. This fruit splits explosively on ripening.

Similar species: Dwarf Spurge is unlikely to be misidentified. Other spurges are larger plants with much broader leaves.

Associated uncommon species: Dwarf Spurge is frequently part of plant communities on calcareous soils that include uncommon species such as Prickly Poppy, Rough Poppy, Venus's-looking-glass, Field Gromwell and Narrow-fruited Cornsalad.

HABITAT:

Dwarf Spurge is found mainly in arable field margins, but also locally in drought-prone grasslands.

SOIL TYPE:

Most frequent on chalky soils, but may also be found on other limestones, calcareous clays and coastal sands.

DISTRIBUTION:

Present throughout southern and eastern England, chiefly on calcareous soils. Restricted to the coasts in the south-west and Wales.

LIFE CYCLE:

Flowers from June to October.
Seed longevity is not known, but may be long.
Germination is mainly in the spring, with smaller numbers germinating in the autumn.

J F M A M J J A S O N D

REASONS FOR DECLINE:

The development of modern crop varieties, the increase in application of nitrogen and the use of broad-spectrum herbicides are likely to be have been detrimental.

Dwarf Spurge: fruit capsule ×8

Sun Spurge

Euphorbia helioscopa

IDENTIFICATION:

Sun Spurge is a hairless plant that can grow up to 50 cm tall. The unusual, yellowish-green flowers are typical of spurges. They lack petals, and are arranged in a loose umbel with 5 rays at the stem apex. Each flower produces a roughly spherical fruit capsule approximately 3–5 mm in diameter. This fruit splits explosively on ripening. The base of the umbel is ringed by a whorl of leaf-like bracts. These bracts are wedge-shaped, increasing in width towards the tip and rarely more than 2 cm wide.

Similar species: Other spurges. The fruits of Broad-leaved Spurge are covered with small warts, and it has serrated leaves arranged alternately up the stem. Petty Spurge is a much-branched plant with leaves arranged alternately along the stem. Dwarf Spurge is generally a much smaller plant with narrow, parallel-sided leaves.

Associated uncommon species: Sun Spurge may be found in a wide range of communities that include uncommon species.

HABITAT:

Sun Spurge can be found in arable field margins, gardens, and other disturbed sites.

SOIL TYPE:

Most frequent on well-drained, rich loamy soils.

DISTRIBUTION:

Present throughout the lowlands of Britain.

LIFE CYCLE:

Flowers from June to October. Seeds are not thought to last for very long in the soil. Germination is throughout the summer from May to September.

J F M A M J J A S O N D

REASONS FOR DECLINE:

Although this species remains frequent over much of the country it has declined in some areas, probably as a result of the increase in winter cropping and the use of broad-spectrum herbicides.

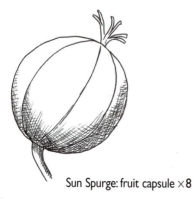

Sun Spurge: fruit capsule ×8

Corn Spurrey

Spergula arvensis

IDENTIFICATION:

A slender, scrambling plant with a stem branching from the base, and growing up to 60cm in length. The leaves are in whorls of four and are very narrow and parallel-sided, up to 3cm long, and covered in glandular hairs. The flowers are in loose, branched clusters at the ends of stems. They are approximately 8mm across with five white petals. The seeds are black and are about 1·5mm in diameter.

Similar species: There are no similar species.

Associated uncommon species: Corn Spurrey can often be found with species such as Corn Marigold, Weasel's-snout and the rarer fumitories.

HABITAT:

Arable and horticultural field margins and other disturbed sites.

SOIL TYPE:

Corn Spurrey is found largely on well-drained sandy soils, but also occurs on clay with flints deposits over chalk.

DISTRIBUTION:

Throughout Britain, but more common in the north and west. It has declined in recent years.

LIFE CYCLE:

Flowers from June to August. Seeds are not thought to last for very long in the soil. Germination is mainly in the spring, but some also occurs in the autumn.

J F M A M J J A S O N D

REASONS FOR DECLINE:

The recent decline is due to a combination of the high levels of nitrogen applied to competitive modern crops, use of broad-spectrum herbicides and increases in the proportion of the cropped area sown with winter cereals.

Corn Spurrey: seed capsule ×4

Musk Stork's-bill

Erodium moschatum

IDENTIFICATION:

Musk Stork's-bill is a hairy plant with spreading stems up to 60 cm long. The leaves are pinnate with toothed or lobed segments, the basal leaves are long-stalked and up to 25 cm long, although the stem leaves are smaller. The leaves have a musky scent when bruised. The clusters of 2–8 flowers are on a long stem that is usually just slightly longer than the basal leaves. The flowers are pinkish-purple, up to 1·5 cm in diameter and longer than the sepals. The stalk of each flower is recurved when the fruits are ripe. Each flower produces five seeds, each of which has a long projection that remains joined to the others (hence 'stork's-bill') until ripe. This projection remains attached to the seed after it is shed from the flower, and helps the seed dig into the ground.

Similar species: The other stork's-bills and crane's-bills are similar. The leaves of stork's-bills are pinnately divided whereas the cranesbills are round or triangular in shape, divided into deep lobes. The leaf segments of Common Stork's-bill are divided to the midrib, those of Musk Stork's-bill only half-way.

Associated uncommon species: On arable land, Musk Stork's-bill often occurs with other uncommon species including Corn Marigold and Weasel's-snout, and in the far west of Cornwall and the Isles of Scilly with Purple Viper's-bugloss and Prickly-fruited Buttercup.

HABITAT:

This plant can be found in several habitats: it occurs in bulb fields on the Isles of Scilly; in the margins of coastal arable fields, and can also be found in Cornish hedges, drought-prone coastal grassland and tracksides.

SOIL TYPE:

Most commonly found on light sandy soils.

MANAGEMENT REQUIREMENTS:

Autumn cultivation.

DISTRIBUTION:

Largely restricted to coastal areas of the south-west of England and Wales, but also found as a casual in waste places elsewhere.

LIFE CYCLE:

Flowers from May to August.

Seed longevity is not known, but may be long.

Germination is likely to be largely in the autumn, but also in the spring in spring-sown crops.

J F M A M J J A S O N D

REASONS FOR DECLINE:

Musk Stork's-bill has maintained its distribution within its restricted range. It is, however, confined to situations that offer the minimum of competition, and the development of modern crop varieties and the increase in amounts of nitrogen applied are likely to be have been detrimental.

Lamb's Succory

Arnoseris minima

Probably extinct

IDENTIFICATION:

A member of the daisy family (Asteraceae), Lamb's Succory grows up to 30 cm tall. The leaves are all in a rosette at the base of the plant and are 5–10 cm in length. They have a few small teeth around the edges and are usually hairless, but are sometimes lightly hairy. The flowers, 7–11 mm in diameter, are compound heads of small, yellow florets produced singly at the ends of erect stems. The stems are inflated and hollow below the flower-heads, and are sometimes simply-branched. The seeds are about 2 mm in length.

Similar species: Lamb's Succory is distinctive but could possibly be confused with other yellow-flowered species in the same family (dandelions, hawkbits and hawkweeds). However, the inflated stem is diagnostic.

HABITAT:

Arable field margins with poor crops, occasionally cultivated land and disturbed tracksides.

SOIL TYPE:

Sands and sandy loams.

Associated uncommon species: Formerly in very species-rich communities.

MANAGEMENT REQUIREMENTS:

Autumn-cultivation.

DISTRIBUTION:

Lamb's Succory was formerly found at scattered localities in sandy areas of the Hampshire Basin, Surrey, Bedfordshire, East Anglia, Lincolnshire and Yorkshire. It has not been recorded since 1971.

LIFE CYCLE:

Flowers from June to August.
Seed lasts for at least 3 years in the soil.
Germination occurs mainly in October and November with a few seedlings emerging in March.

J F M A M J J A S O N D

REASONS FOR DECLINE:

A major cause has been the development of high-yielding crop varieties and the application of large quantities of nitrogen. The marginal arable sites that Lamb's Succory formerly occupied have also been susceptible to abandonment.

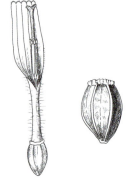

Lamb's Succory: floret (*left*) ×2, seed (*right*) ×2

Slender Tare

Vicia parviflora

IDENTIFICATION:

Slender Tare is a scrambling species growing up to 60 cm in length. The leaves are divided into 2–5 pairs of leaflets, each of which is up to 2·5 cm long. The flowers are pale purple and are 6–8 mm long. They are borne in groups of 2–8 on stalks longer than the leaves. Each flower produces a 5–8 seeded pod about 3 mm in width.

Similar species: Hairy Tare and Smooth Tare, which are both scrambling vetches with small flowers and narrow leaves, are similar to Slender Tare. The key identification features of the three species are as follows:

SLENDER TARE: pale purple flower 6–8 mm; seed pod with 5–8 seeds.

HAIRY TARE: white/purplish flower 4–5 mm; seed pod with 2 seeds.

SMOOTH TARE: pale blue flower *ca.* 4 mm; seed pod with 4 seeds.

Associated uncommon species: On arable land, it can occur with species including Corn Buttercup, Spreading Hedge-parsley, Shepherd's-needle, Broad-leaved Spurge and, sometimes, Yellow Vetchling.

HABITAT:

Arable field margins, sunny hedge-banks, tracksides, road verges, walls and clay cliffs. In Somerset also on cultivated ground, including flower-beds.

SOIL TYPE:

Calcareous clay loams; less commonly on chalk.

DISTRIBUTION:

Slender Tare occurs in the south Midlands, Somerset, Dorset Oxfordshire and Essex. It has become much rarer within this range in recent years.

LIFE CYCLE:

Flowers from June to August. Seed longevity is unknown. Germination is probably mainly in the autumn.

J F M A M J J A S O N D

REASONS FOR DECLINE:

Reasons for the decline are unknown, but are probably the increases in use of nitrogen and susceptibility to broad-spectrum herbicides.

Slender Tare: seed pod × 3

Thorow-wax

Bupleurum rotundifolium

IDENTIFICATION:

This is a very distinctive member of the carrot family (Apiaceae) that grows up to 75 cm tall, with a reddish, branched stem. The leaves are round to oval, the uppermost surrounding the stem. The greenish-yellow flowers have small yellowish petals and are gathered in clusters at the end of stems. Each cluster is surrounded by a ring of green, petal-like bracteoles. Each flower produces a pair of brown seeds, each seed measuring 3·0–3·5mm.

Similar species: False Thorow-wax sometimes occurs as a bird-seed alien. It is very similar to Thorow-wax, but with narrowly oval leaves.

Associated uncommon species: Thorow-wax grew with Pheasant's-eye on at least one site before its demise.

HABITAT:

Arable field margins, but now only as a casual of disturbed and waste ground.

SOIL TYPE:

Calcareous loams.

MANAGEMENT REQUIREMENTS:

Autumn cultivation.

E?

DISTRIBUTION:

Formerly scattered throughout central-southern England, with isolated sites as far north as Yorkshire. It was common in the 18th and 19th centuries, but it has not been seen as a native plant in Britain since 1973.

LIFE CYCLE:

Flowers from June to July.
Seed is short-lived in the soil.
Germination is mainly in October and November, but some seedlings appear in April.

J F M A M J J A S O N D

REASONS FOR DECLINE:

Some of the decline can probably be attributed to improved seed cleaning at the end of the 19th century. Short-lived seed makes this species very sensitive to changes in management. It may be susceptible to herbicides, but seems to compete quite well with a highly fertilised crop.

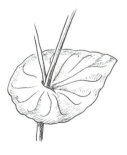

Thorow-wax: fruit (*left*) ×8, leaf (*right*) × 1

Small Toadflax

Chaenorhinum minus

IDENTIFICATION:

Small Toadflax is an upright, branched plant, rarely growing to more than 20 cm tall. The leaves are narrow and parallel-sided, arranged alternately. The solitary flowers have short stalks and are borne in the axils of leaves. The flowers have bilateral symmetry. They are purple outside, paler inside, 6–9 mm long and with a short spur. The seed capsule is oval, and opens at the top.

Similar species: Small Toadflax resembles other toadflaxes and Weasel's-snout in general form, although the flowers are distinctive.

Associated uncommon species: Small Toadflax is frequently found in species-rich arable plant communities on chalk or limestone soils with species including Rough Poppy, Night-flowering Catchfly, Field Gromwell and Pheasant's-eye.

HABITAT:

Arable field margins, horticultural land, tracks and railway ballast.

SOIL TYPE:

Mainly found on light calcareous loams.

DISTRIBUTION:

Small Toadflax is widespread throughout lowland England, Wales and southern Scotland.

LIFE CYCLE:

Flowers from June to July, and also in stubbles after harvest.

Seed is thought to be relatively short-lived.

Germination is almost entirely in the spring.

J F M A M J J A S O N D

REASONS FOR DECLINE:

Small Toadflax is probably susceptible to many herbicides, and is a non-competitive species that does not grow well in a fully-fertilised modern crop variety. Because it also flowers in stubbles is has probably been affected by early ploughing.

Small Toadflax: seed capsule ×4

Smaller Tree-mallow

Lavatera cretica

IDENTIFICATION:

Smaller Tree-mallow can be either annual or biennial and grows up to 1·5 m tall. The lower parts of the stem have numerous short, star-shaped hairs. The leaves are shallowly 5–7 lobed. The flowers have 5 pink-lilac petals approximately 15 mm long. The seeds are smooth and yellowish.

Similar species: Other mallows. *Lavatera* species can be distinguished from the more common *Malva* species, Common Mallow and Dwarf Mallow, by differences in the outer ring of sepals. *Lavatera* has three outer sepals while *Malva* has 6–9. The familiar garden plant *Lavatera trimestris*, with large, bright pink flowers, is occasionally naturalised.

Associated uncommon species: On the Isles of Scilly it grows with Small-flowered Catchfly and Western Ramping-fumitory.

HABITAT:

Bulb-fields, field borders, roadsides and other disturbed ground

SOIL TYPE:

Sandy, stony loams.

MANAGEMENT REQUIREMENTS:

In bulb-fields, field margins should be left free from herbicide applications.

DISTRIBUTION:

Established colonies occur only on the Isles of Scilly where it is locally frequent, although it also occurs as a casual elsewhere.

LIFE CYCLE:

Flowers from June to July.
Seed longevity unknown.
Germination is probably in the autumn.

J F M A M J J A S O N D

REASONS FOR DECLINE:

Smaller Tree-mallow has declined in the past with increased herbicide use in bulb-fields. However, many populations have recently recovered.

Smaller Tree-mallow: fruit × 3

Venus's-looking-glass
Legousia hybrida

IDENTIFICATION:
Venus's-looking-glass grows between 3–25 cm tall. The upper leaves are oval with wavy edges, 1·5 cm long and lack stalks. The stem has short hairs. The 5-petalled flowers are up to 10 mm across when fully open, but are often closed when the sun is not shining. The petals are normally purple, but can be lilac or even white. The sepals are longer than the petals. The fruit capsule measures 30 mm × 4 mm and is a very distinctive, long, almost cylindrical, shape. It is present throughout the flowering period.

Similar species: Greater Venus's-looking-glass is a very rare species that is completely hairless, with much larger purple flowers.

Associated uncommon species: Venus's-looking-glass grows in a number of different plant communities, often with uncommon species including Rough and Prickly Poppies, Shepherd's-needle and Narrow-fruited Cornsalad.

HABITAT:
Arable fields.

SOIL TYPE:
Calcareous loams, calcareous clay loams, sometimes on calcareous sands.

DISTRIBUTION:
Venus's-looking-glass is widespread in eastern and central-southern England, but also occurs as far north as the North York Moors. Although locally common in the south-east, it is rare in the west and absent from Wales. It has disappeared from many western and midland sites.

LIFE CYCLE:
Flowers from May to August. Seed is thought to be long-lived. Germination is in both the spring and autumn.

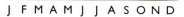

J F M A M J J A S O N D

REASONS FOR DECLINE:
It is very susceptible to many herbicides, but is quite shade tolerant. It can persist under a dense crop canopy more than many other low-growing species.

Venus's-looking-glass: flower × 1

Greater Venus's-looking-glass

Legousia speculum-veneris

IDENTIFICATION:

Greater Venus's-looking-glass grows from 10–40 cm tall. The upper leaves are oval with wavy edges and lack stalks. The stem is hairless. The 5-petalled flowers are up to 25 mm across when fully open, but are often closed when the sun is not shining. The petals are normally purple, but occasionally can be lilac or even white. The sepals are as long as the petals. The fruit capsule is present throughout the flowering period but is shorter than that of Venus's-looking-glass. It measures up to 15 mm in length.

Similar species: Venus's-looking-glass is a much more frequent species. Its flowers are much smaller, the fruit capsule larger, and the stem hairy.

Associated uncommon species: Unknown

SOIL TYPE:

Calcareous loams.

DISTRIBUTION:

A very rare species of the south of England. Distribution very poorly known, but known since 1916 from one site in Hampshire. More common than Venus's-looking-glass in much of mainland Europe.

LIFE CYCLE:

Flowers from March to June. Seed is probably long-lived. Germination is mainly in the autumn.

J F M A M J J A S O N D

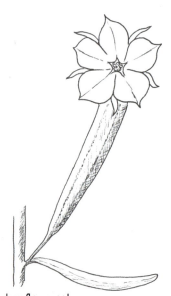

Greater Venus's-looking-glass: flower × 1

Yellow Vetchling

Lathyrus aphaca

IDENTIFICATION:

Yellow Vetchling has yellow flowers typical of the pea family (Fabaceae). The flowers are borne singly on long stems in the leaf axils, and are approximately 10 mm long. True leaves are absent in mature plants, but the stipules are enlarged to take the place of leaves. These are large and triangular, and are borne in pairs along the stem, with a flower and a tendril at each node. Seeds are enclosed in a pod 2–3 cm in length.

Similar species: Meadow Vetchling also has yellow pea-like flowers but has true leaves with a single pair of leaflets and clusters of 5–12 flowers.

Associated uncommon species: In arable edge habitats, often with uncommon species such as Rough Poppy, Narrow-fruited Cornsalad and Shepherd's-needle.

HABITAT:

Occasionally found in cultivated field margins, but also occurs on road verges, clay cliffs and in disturbed calcareous grassland.

SOIL TYPE:

Calcareous loams and calcareous clay loams.

MANAGEMENT REQUIREMENTS:

Biennial cultivation in the autumn.

DISTRIBUTION:
Most frequent on limestone near the coast in southern England, but also inland as far north as Cambridgeshire.

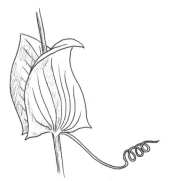

Yellow Vetchling: leaf-like stipule × 1

LIFE CYCLE:
Flowers from June to August.
Seed longevity is unknown.
Germination is mainly in the autumn.

J F M A M J J A S O N D

REASONS FOR DECLINE:
Formerly introduced as a contaminant of legume crops, its decline may be due to improved seed-cleaning techniques.

Purple Viper's-bugloss

Echium plantagineum

IDENTIFICATION:

Purple Viper's-bugloss grows up to 75 cm tall, and is covered with bristly hairs. The leaves are narrowly oval, up to 15 cm long and 3 cm wide, forming a rosette at the base of the stem but also occurring along the stem itself with the upper leaves clasping the stem. The flowers are large, approximately 2 cm long, with petals fused into a tube with two lips at the opening, and two projecting stamens. The flowers are arranged in a spike and are light purple with darker streaks, turning dark blue with age. The seeds are 2·5 mm long, pyramidal in shape, pitted, warty and hard.

Similar species: The only similar species is Viper's-bugloss, which occasionally occurs in chalky field edges. Viper's-bugloss differs in having flowers with 4–5 projecting stamens.

Associated uncommon species: Corn Marigold, Weasel's-snout and Musk Stork's-bill are present at the remaining site.

HABITAT:

Isles of Scilly bulb fields, arable fields and associated hedgerow-bottoms.

SOIL TYPE:

The only mainland British site is on sandy loams over slate.

MANAGEMENT REQUIREMENTS:

Purple Viper's-bugloss appears to thrive under a regime of spring cereals and rotational set-aside. Crops are not very competitive at the remaining site.

DISTRIBUTION:
Purple Viper's-bugloss has only ever been established in Britain in the far west of Cornwall and the Isles of Scilly. It is now known only from a few fields near Land's End, where numbers can be very large. It still occurs as a casual elsewhere.

LIFE CYCLE:
Flowers from June until harvest, and again in the stubbles until killed by frost.
Seed appears to be long-lived. Germination can occur from autumn to spring.

J F M A M J J A S O N D

REASONS FOR DECLINE:
Purple Viper's-bugloss has never been common. Its present distribution is restricted by the abandonment of arable land around the Cornish coast.

Purple Viper's-bugloss:
leaf ×0·5 showing soft down and prominent lateral veins on underside

Weasel's-snout OR Lesser Snapdragon

Misopates orontium

IDENTIFICATION:

Weasel's-snout has an erect, branched, leafy stem, growing up to 50 cm tall but usually less. The shiny, dark-green leaves are very narrowly oval and up to 5 cm long. The stalkless flowers are borne in the axils of the upper leaves. These are deep pink (rarely white), and up to 15 mm across. They have bilateral symmetry and resemble those of the garden snapdragon. The fruit capsule is oval, approximately 1 cm in length, opening with three holes at the apex.

Similar species: No other species found in arable habitats resembles Weasel's-snout.

Associated uncommon species: Weasel's-snout usually grows with Corn Marigold, and sometimes with Small-flowered Catchfly and Lesser Quaking-grass.

HABITAT:

Arable field margins, especially root crops. Also in gardens, allotments and other disturbed places.

SOIL TYPE:

Sandy loams, occasionally stony clays in south-west England.

MANAGEMENT REQUIREMENTS:

Spring cultivation.

DISTRIBUTION:

Confined to the south of Britain, and most frequent near the coasts of the south and west from Hampshire to Cornwall and Pembrokeshire. It has declined greatly in the eastern part of its range, but is still quite widespread in the west, with isolated recent records elsewhere.

LIFE CYCLE:

Flowers from late June to October. Seed is long-lived.

Germination is mainly from March to May, although there is some during the summer and autumn.

J F M A M J J A S O N D

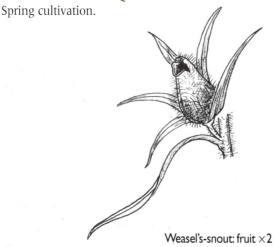

Weasel's-snout: fruit ×2

REASONS FOR DECLINE:

Susceptible to many broad-spectrum herbicides, growing poorly in fully fertilised modern crops. Also affected by the declining area of spring-sown crops, especially in eastern England, and the conversion of arable land to pasture in the West.

Field Woundwort

Stachys arvensis

IDENTIFICATION:

Field Woundwort has weak, spreading, but sometimes erect, stems that are branched from the base. It rarely grows more than 20 cm tall. The leaves are oval to heart-shaped, with irregularly toothed edges. They are hairy, have short stalks and are arranged on the stem as opposite pairs. The flowers are arranged in whorls of up to 6 at the ends of the stems and form a loose spike. The pale purple flowers have bilateral symmetry typical of the mint family (Lamiaceae) and are up to 7 mm long, with a large lower lip and three smaller lobes. Each flower forms four seeds.

Similar species: Other species in the mint family may be confused with this species. The Corn Mint is the most similar species found in arable land, but this has radially symmetrical flowers and a characteristic light scent of mint.

Associated uncommon species: Field Woundwort can frequently be found with Corn Marigold, Weasel's-snout and other species of acidic sandy soils in the west of Britain.

HABITAT:

Found largely in arable field margins and occasionally in gardens.

SOIL TYPE:

Most frequent on non-calcareous soils. It is characteristic of sandy loam soils, but can also be found on clay with flint deposits over chalk.

DISTRIBUTION:

Present throughout the lowlands of Britain, but very local in Scotland. It is most common in the west of the country.

LIFE CYCLE:

Flowers from April to November. Little is known about seed biology, but it is likely to be quite long-lived. Germination is mainly in the spring.

J F M A M J J A S O N D

REASONS FOR DECLINE:

Although this species remains frequent over much of its range it has declined in some areas, probably as a result of the increase in winter cropping, the use of broad-spectrum herbicides and increases in the amount of nitrogen applied to competitive modern crop varieties.

Field Woundwort: flower ×5

Identification key to arable plants

The following section is an identification key to the arable plants featured in this book, along with any species with which they may be confused.

How to use the key
The key starts with an introduction to plant famiⅼes and provides information on the important features in the identification of species from that family.

The key is ordered by flower characteristics (number of petals, shape, size and colour). The information realting to each species is presented as shown below:

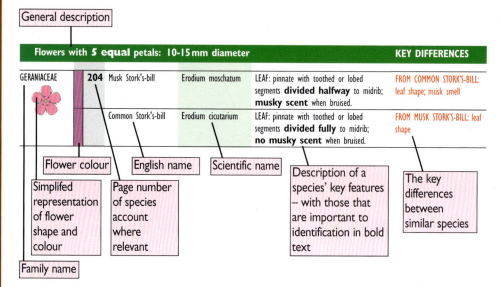

In each of the flower categories details of the key identification features of each plant are given along with, in the case of plants featured in the book, the relevant page number where a full description of the plant can be found.

AN INTRODUCTION TO PLANT FAMILIES

FAMILY	DESCRIPTION	IMPORTANT IDENTIFICATION FEATURES
Ranuculaceae **BUTTERCUP**	Recognised by warm, shiny 5-petalled yellow flowers; leaves generally dissected — wedge-shaped or 3-lobed; hairs variable. Genera include: Adonis, Consolida, Myosurus, Ranunculus.	Flower size, fruit or position of sepals.
Papaveraceae **POPPY**	Scarlet/orange, 4-petalled single flowers with slightly crumpled appearance.	Fruit capsules diagnostic.
Fumariaceae **FUMITORY**	Plants of cultivated or disturbed ground which gained their name from the French for burnt earth, Fume terre, as they look scorched in appearance; thin stems, hairless and 2-lipped flowers form short racemes; alternate leaves which are much-divided.	Rounded fruit, flower size and position and shape of sepals.

AN INTRODUCTION TO PLANT FAMILIES

FAMILY	DESCRIPTION	IMPORTANT IDENTIFICATION FEATURES
Urticaceae **NETTLE**	Known for stinging hairs; male and female flowers on different plants.	Distinctive and well-known plant.
Chenopodiaceae **GOOSEFOOT**	Flowers small and greenish; leaves flat and mealy looking; plants formerly used as food-plants. Genera include: Chenopodium.	Leaf shape.
Caryophyllaceae **PINK**	4–5 petals which may be un-notched, notched or dissected; leaves nearly always opposite at right-angles to the adjacent ones. Genera include: Agrostemma, Ceratstium, Silene, Spergula, Spergularia.	Different genera within this group can be told by looking at the styles which number from 2–5.
Polygonaceae **KNOTWEED**	Leaves alternate with flowers in small cluster. Genera include: Fallopia, Polygonum.	Sheath at base of leaves is an ochrea (pl. ochreae). The size, shape and colour of these are diagnostic.
Malvaceae **MALLOW**	Woody plants; leaves are palmately-lobed and very soft. Pink/mauve showy flowers have five petals. Genera include: Lavatera.	Fruit and sepal characteristics.
Violaceae **VIOLET**	Flowers pansy-shaped; five sepals and petals plus spur. Genera include: Viola.	Flower characteristics, leaf shape and spur.
Brassicaceae **CABBAGE**	4 white or yellow petals arranged in the shape of a cross; many plants in this family taste of mustard. Some species have unequal petals e.g. Wild Candytuft. Genera inlcude: Alyssum, Arabis, Capsella, Coronopus, Diplotaxis, Iberis, Raphanus, Teesdalia, Thlaspi.	Fruit shape diagnostic - either long and thin (siliqua) or broad and short (silicula).
Primulaceae **PRIMROSE**	5-petalled flowers; narrow pointed sepals.	Flower colour and sepal length.
Fabaceae **PEA**	Flowers pea-shaped comprising upper petals and wings and the lower two fused to form keel; leaves pinnate, terminating in a tendril.	Flower colour; shape of seed pod and number of seeds.
Lythraceae **LOOSESTRIFE**	Simple leaves; 6 pink petals and 4–6 stamens which do not protrude.	Distinctive.
Euphorbiaceae **SPURGE**	Plants contain white milky substance (latex); leaves are simple. The herbalist Gerard compared the flower-head to that of a juggler's hands.	The distinctive, specialised flower head found in spurges whih is made up of glands and cyathium.
Geranaceae **CRANE'S-BILL**	Leaves deeply lobed — hand-like in Geranium and pinnate in Erodium; flowers 5-petalled with prominent stamens.	Fruit resembles bill of stork.
Apiaceae **CARROT**	Heads of flowers resemble an umbrella and originally provided many of our vegetables and herbs. NB Some members of this family are poisonous. Flowers vary and can be white, tinged pink or yellow-green. Genera include: Aethusa, Anthriscus, Apium, Bunium, Bupleurum, Petroselinum, Scandix, Torilis.	Leaves, fruits, presence or absence of bracts and time of appearance of the plant.
Solanaceae **POTATO**	Twining plant, flowers greenish-white, leaves heart shaped.	Inconspicuous flower; seed characteristics; twining.
Convolvulaceae **BINDWEED**	Twining plants with heart-shaped leaves and white or pink funnel-shaped flowers.	Large flowers; leaf shape; twining.
Boraginaceae **BORAGE**	Flowers often unfurl on one side, changing colour from pink to blue; bristly stems and leaves. Genera include: Anchusa, Buglossoides, Cynoglossum Echium, Lithospermum, Myosotis.	Leaf and flower characteristics.
Lamiaceae **DEAD-NETTLE**	Characteristic plants which have square stems, paired leaves and lipped flowers. Mostly scented. Genera: Lamium, Mentha, Stachys.	Shape of leaf, flower details — colour, spots or position on plant.
Scrophulariaceae **FIGWORT**	Stems often square; flowers often have a two-lipped corolla; leaves opposite or alternate. Genera: Antirrhinum, Chaenorrhinum, Kickxia, Misopates, Veronica.	Variable group; the important features depend on the genus.
Campanulaceae **BELLFLOWER**	Bell-shaped flowers.	Flower and fruit characteristics.
Rubiaceae **BEDSTRAW**	Scrambling plants with square stem with leaves in whorls around it; 4-petalled flowers in whorls.	Shape and location of seed.
Valerianaceae **VALERIAN**	Small flowers borne in clusters. Genera: Valeriana, Valerianella.	Fruits are diagnostic.
Asteraceae **DAISY**	A large group in which the flower is composed of an inflorescence which may have disc florets, ray florets (Dandelion) or both (e.g. Daisy). Genera: Anthemis, Bellis, Centaurea, Chamomilla, Chrysanthemum, Cynara, Filago, Hypochaeris.	Variable group; the important features depend on the genus and include: flower shape and size; position of leaves; and whether or not the plant contains latex.

EUPHORBIACEAE	196	Broad-leaved Spurge	Euphorbia platyphyllos	LEAF: narrowly oblong; pointed at tip; serrated edge; arranged alternately up stem. FRUIT: seed capsule nearly spherical; covered in small warts; 3 mm wide.	FROM PETTY SPURGE AND SUN SPURGE: serrated leaves and warty fruits
		Petty Spurge	Euphorbia peplus	HEIGHT: 10—30 cm. LEAF: oval, broader towards tip; 5—30 mm long, arranged alternately on stem. FRUIT: seed capsule rounded, triangular; 2 mm wide.	much-branched
	198	Dwarf Spurge	Euphorbia exigua	LEAF: narrow, parallel-sided. FRUIT: seed capsule nearly spherical; smooth; 2 mm wide.	small size
	200	Sun Spurge	Euphorbia helioscopia	LEAF: wedge-shaped, broader towards tip; whorl of leaf-like bracts below umbel. FRUIT: seed capsule seed capsule nearly spherical; smooth; 3—5 mm wide.	FROM BROAD-LEAVED SPURGE AND DWARF SPURGE: whorl of leaf-like bracts; wartless fruits
POLYGONACEAE	58	Black-bindweed	Fallopia convolvulus	LEAF: green, heart-shaped; 2—6 cm long. SEED: triangular and black 5 mm long.	flowers very inconspicuous in comparison with CONVOLVULACEAE (page 234); twines anti-clockwise

BRASSICACEAE	96	Thale Cress	Arabidopsis thaliana	STEM LEAVES: small and very narrow. SEED POD: long, thin and curved slightly upwards; 10—18 mm long.	shape of seed pod distinctive
		Shepherd's-purse	Capsella bursa-pastoris	HEIGHT: 3—40 cm; from basal rosette. STEM LEAVES: spear-shaped. SEED POD: inverted, notched triangle shape; 6—9 mm wide.	shape of seed pod distinctive
		Shepherd's Cress	Teesdalia nudicaulis	HEIGHT: 8—45 cm; from basal rosette. STEM LEAVES: absent from central stem. SEED POD: spoon-shaped; 4 mm wide.	shape of seed pod distinctive
	166	Perfoliate Penny-cress	Thlaspi perfoliatum	STEM LEAVES: oval with basal lobes clasping stem. FLOWER: 2·0—2·5 mm wide. FRUIT: heart-shaped; central part entirely circled by broad wings; on stalks parallel to the ground; 5 mm long by 4 mm wide.	FROM FIELD PENNY-CRESS: fruit shape distinctive (left)
	164	Field Penny-cress	Thlaspi arvense	STEM LEAVES: oblong with auricles clasping stem. FLOWER: 4—6 mm wide. FRUIT: circular; central part entirely circled by broad wings; on upwardly curving stalks; 12—20 mm wide.	FROM PERFOLIATE PENNY-CRESS: fruit shape distinctive (left)
	70	Wild Candytuft	Iberis amara	LEAVES: lower spoon-shaped, upper spear-shaped. FLOWER: petals of unequal size. FRUIT: round with triangular projecting wings; 4—5 mm wide.	characteristic flower and fruit shape
	54	Small Alison	Alyssum alyssoides	LEAVES: spear-shaped with star-like hairs giving a grey appearance. FLOWER: fades to white over time. FRUIT: spherical; 3 mm wide.	combination of yellow flowers and star-shaped hairs diagnostic
		Flixweed	Decurania sophia	HEIGHT: 30—80 cm. STEM LEAVES: pinnately divided. SEED POD: long, thin and curved slightly upwards; 15—25 mm long.	tall plant; yellow flowers; pinnate leaves

Flowers with 4 equal petals: <10mm diameter | KEY DIFFERENCES

RUBACEAE	80	Corn Cleavers	Galium tricornutum	PLANT: pale green; scrambling; stems sharply 4-angled; up to 80 cm long. LEAVES: narrow, arranged in whorls of 6—8; up to 50mm long by 5 mm wide. FLOWER: greenish-white; in groups of 3; 1·0—1·5 mm wide. FRUIT: spherical; covered in papillae; on strongly recurved stalks; 3mm wide.	FROM CLEAVERS AND FALSE CLEAVERS: seed stalks recurved; seeds covered in papillae
		Cleavers	Galium aparine	PLANT: dark green; scrambling; stems sharply 4-angled; up to 120 cm long. LEAVES: narrow, arranged in whorls of 6—8; up to 50mm long by 2—3 mm wide. FLOWER: greenish white; in groups of 2—5; 4—8 leaf-like bracts; 2mm wide. FRUIT: spherical; green or purplish; covered in hooked bristles on straight stalk; 4—6mm wide.	seed stalks straight; seeds with hooked bristles
		False Cleavers	Galium spurium	PLANT: dark green; scrambling; stems sharply 4-angled; up to 60 cm long. LEAVES: narrow, arranged in whorls of 6—8; 5—18mm long. FLOWER: greenish; in groups of 3—9; 2—3 leaf-like bracts; 1 mm wide. FRUIT: spherical; blackish; covered in hooked bristles on straight stalk; 1·5—3 mm wide.	seed stalks straight; seeds with hooked bristles
	152	Field Madder	Sherardia arvensis	PLANT: prostrate, mat forming; 5—40 cm. LEAVES: lower leaves broadly spear-shaped in whorls of 4; upper leaves narrow; prickly-edged in whorls of 6. FLOWER: lilac; in dense short-stalked clusters of 4—10; 2—3mm wide.	FROM OTHER CLEAVERS: flower colour and four-leaved whorl of lower leaves diagnostic

Flowers with 4 ± equal petals: >10mm diameter | KEY DIFFERENCES

PAPAVERACEAE	174	Babbington's Poppy	Papaver dubium ssp. lecoquii	HEIGHT: up to 60 cm. FLOWER: pinkish-red, unblotched; persist usually only for a single day; anthers yellow to brown; 50 mm wide. SEED CAPSULE: elongated; without prickles; 20—25mm long.	seed capsule bristle-free (left); FROM LONG-HEADED POPPY: latex exuded from stem turns yellow on contact with the air
	176	Common Poppy	Papaver rhoeas	HEIGHT: up to 80 cm. FLOWER: bright scarlet often blotched black at base; anthers blue-black; 80 mm wide. SEED CAPSULE: seed capsule spherical with flattened top; without prickles; 10—20mm long.	seed capsule distinctive (left)
	178	Long-headed Poppy	Papaver dubium	HEIGHT: up to 80 cm. FLOWER: pinkish-red, unblotched; persist usually only for a single day; anthers brown to blue-black; 50 mm wide. SEED CAPSULE: elongated; without prickles; 20—25mm long.	seed capsule bristle-free (left); FROM BABBINGTON'S POPPY: latex exuded from stem turns white on contact with the air
	180	Prickly Poppy	Papaver argemone	HEIGHT: up to 50 cm. FLOWER: orange-red, unblotched; persist usually only for a single day; 50 mm wide. SEED CAPSULE: seed capsule long and narrow; with prickles; 20—25mm long.	seed capsule distinctive (left)
	182	Rough Poppy	Papaver hybridum	HEIGHT: up to 50 cm. FLOWER: scarlet, black blotches at base; persist usually only for a single day. SEED CAPSULE: spherical; with prickles; 10—15mm wide.	seed capsule distinctive (left)

SCROPHULARIACEAE

				KEY DIFFERENCES
186	Breckland Speedwell	Veronica praecox	PLANT: erect, sometimes branched. LEAF: toothed but not deeply divided. FLOWER: blue, streaked dark blue; longer than surrounding calyx; 3 mm wide; long stalk. FRUIT: capsule bi-lobed; longer than broad as long; seeds cup-shaped.	leaf shape; flower colour
188	Fingered Speedwell	Veronica triphyllos	PLANT: erect, sometimes branched. LEAF: divided into 3 parallel lobes; upper leaves stalkless. FLOWER: deep blue; shorter than surrounding calyx; 3—4 mm wide; long stalk. FRUIT: capsule deeply bi-lobed; as broad as long; seeds cup-shaped.	leaf shape; flower colour
190	Spring Speedwell	Vernonica verna	PLANT: erect, sometimes branched. LEAF: pinnately lobed with 5—7 segments. FLOWER: sky blue; 2—3 mm wide; short stalk. FRUIT: capsule bi-lobed; broader than long.	leaf shape; flower colour
	Wall Speedwell	Veronica arvensis	PLANT: erect, sometimes branched. LEAF: toothed leaves. FLOWER: blue; shorter than surrounding calyx; 1 mm wide; without stalk. FRUIT: capsule bi-lobed; as broad as long; seeds flat.	leaf shape; flower colour
192	Green Field-speedwell	Veronica agrestis	PLANT: trailing, mat-forming; much-branched. LEAF: oval; green; toothed edges; 5—15 mm across. FLOWER: pale blue with white lower lobe and centre; 5 mm wide; flower stalk shorter than leaves. FRUIT: bi-lobed with erect lobes; broader than long.	seed shape diagnostic (left)
194	Grey Field-speedwell	Veronica polita	PLANT: trailing, mat-forming; much-branched. LEAF: oval; greyish-green; 5—15 mm across. FLOWER: bright blue with lower lobe occasionally paler blue; 5 mm wide; flower stalk shorter than leaves. FRUIT: bi-lobed with erect lobes; broader than long.	seed shape diagnostic (left)
	Common Field-speedwell	Veronica persica	PLANT: trailing, mat-forming; much-branched. LEAF: oval; dark green; 10—30 mm across. FLOWER: bright blue with lower lobe sometimes paler blue or white; 8—12 mm wide; flower stalk longer than leaves. FRUIT: bilobed with spreading lobes; 2x as broad as long.	seed shape diagnostic (left)

VALERIANACEAE

				KEY DIFFERENCES
86	Broad-fruited Cornsalad	Valerianella rimosa	FLOWER: white-pinkish. SEED CAPSULE: distinctive shape; 1·5 mm wide.	seed shape diagnostic (left)
88	Common Cornsalad	Valerianella locusta	FLOWER: white-pale blue. SEED CAPSULE: distinctive shape; 1·8—2·5 mm wide.	seed shape diagnostic (left)
90	Keeled-fruited Cornsalad	Valerianella carinata	FLOWER: white-pale blue. SEED CAPSULE: distinctive shape; 0·8—1·4 mm wide.	seed shape diagnostic (left)
92	Narrow-fruited Cornsalad	Valerianella dentata	FLOWER: white-pinkish. SEED CAPSULE: distinctive shape; 0·75 mm wide.	seed shape diagnostic (left)

Flowers with 5 equal petals: 2–8 mm diameter | KEY DIFFERENCES

					KEY DIFFERENCES
POLYGONACEAE	148	Cornfield Knotgrass	Polygonum rurivagum	PLANT: scrambling-erect, branched; to 30 cm. LEAVES: long and thin; grey-green; ochreae brownish-red below; leaf stalks enclosed in ochreae; 15–35 mm long by 2–4 mm wide; ochreae 10 mm long. FLOWER: pinkish; 1–2 clustered. FRUIT: projects slightly from flower; 3·0 mm × 1·5 mm.	Leaf shape and size; long ochreae; fruit projection
		Knotgrass	Polygonum aviculare	PLANT: erect-spreading; to 200 cm. LEAVES: long oval; green; ochreae silvery-white, brownish at base; leaf stalks enclosed in ochreae; 25–50 mm long by 5–15 mm wide; ochreae 5 mm long. FLOWER: pink to greenish-white; 1–2 clustered. FRUIT: enclosed within flower; 3-sided, all equal; 3·0 mm × 1·5 mm.	leaf shape and size; fruit shape
		Northern Knotgrass	Polygonum boreale	PLANT: erect-spreading; to 100 cm. LEAVES: spoon-shaped; green; ochreae silvery or brownish; leaf stalks project from ochreae; 30–50 mm long by 5–18 mm wide; ochreae 5 mm long. FLOWER: white with pinkish edges; 1–2 clustered. FRUIT: enclosed within flower; 3-sided, all broad; 4·0 mm × 2·5 mm.	leaf shape and size; leaf stalk projecting from ochreae; fruit shape
		Equal-leaved Knotgrass	Polygonum arenastrum	PLANT: prostrate, mat-forming; to 30 cm. LEAVES: round-oval; hairless; green; ochreae reddish-brown; leaf stalks enclosed in ochreae; 20 mm long by 5 mm wide; ochreae 5 mm long. FLOWER: greenish-white or pink; 2–3 clustered FRUIT: enclosed within flower; 3-sided, two broad and one narrow; 1·5 mm × 2·5 mm.	leaf shape and size; flower clusters; fruit shape
CARYOPHYLLACEAE	56	Four-leaved Allseed	Polycarpon tetraphyllum	PLANT: prostrate; to 15 cm. LEAVES: oval, **arranged in whorls**. FLOWER: white; 2–3 mm wide.	whorled arrangement of leaves diagnostic. The similar chickweeds, and mouse-ears do not have whorled leaves.
	202	Corn Spurrey	Spergula arvensis	PLANT: scrambling, slender; stem branching from the base; to 60 cm. LEAVES: narrow, parallel-sided; in whorls of 4. FLOWER: white; not notched; in loose, branched clusters.	unmistakable

Flowers with 5 equal petals: 10–12 mm diameter | KEY DIFFERENCES

					KEY DIFFERENCES
PRIMULACEAE	172	Blue Pimpernel	Anagallis arvensis ssp. foemina	FLOWER: blue; narrow petals **do not** overlap; stalks **shorter** than leaves; sepals **longer** than petals.	petals do not overlap; sepal length; flower stalk length
	172	Scarlet Pimpernel (blue-flowered form)	Anagallis arvensis	FLOWER: blue or lilac; petals overlap; stalks **longer** than leaves; sepals **shorter** than petals.	petals overlap; sepal length; flower stalk length
	172	Scarlet Pimpernel	Anagallis arvensis	FLOWER: orange-red or pink; stalks longer than leaves; sepals shorter than petals. LEAF: underside covered in black dots.	flower colour

GERANIACEAE	204	Musk Stork's-bill	Erodium moschatum	LEAF: pinnate with toothed or lobed segments **divided halfway** to midrib; **musky scent** when bruised.	FROM COMMON STORK'S-BILL: leaf shape; musk smell
		Common Stork's-bill	Erodium cicutarium	LEAF: pinnate with toothed or lobed segments **divided fully** to midrib; **no musky scent** when bruised.	FROM MUSK STORK'S-BILL: leaf shape

CARYOPHYLLACEAE	74	Night-flowering Catchfly	Silene noctiflora	PLANT: erect; hairy; with sticky stems. FLOWER: slightly notched; white-pink inside, creamy yellow on backs; 20 mm wide.	flowers tightly closed by mid-morning
	76	Small-flowered Catchfly	Silene gallica	PLANT: erect; hairy; with sticky stems. FLOWER: slightly notched; white or pink; 15 mm wide.	FROM NIGHT-FLOWERING CATCHFLY: flowers open all day
		White Campion	Silene latifolium	PLANT: erect; hairy; upper stem **not sticky**. FLOWER: lightly notched; white; 25–30 mm wide.	FROM CATCHFLY SPP.: large flowers; upper stem not sticky
	82	Corncockle	Agrostemma githago	PLANT: erect; hairy; upper stem not sticky. FLOWER: trumpet-shaped; bright pink; 35 mm wide; long pointed sepals	unmistakable
MALVACEAE	214	Smaller Tree-mallow	Lavatera cretica	PLANT: erect; lower part of stem with stiff, star-shaped hairs; up to 150 cm. LEAF: shallowly 5–7-lobed. FLOWER: petals **barely** overlapping; 3 outer sepals. SEED: yellowish; smooth; rounded angle.	stems not woody; smaller flowers with petals barely overlapping
		Tree-mallow	Lavatera arborea	PLANT: erect; lower part of stem woody, upper softly hairy; up to 300 cm LEAF: shallowly 5–7-lobed. FLOWER: petals overlapping; 3 outer sepals. SEED: yellowish; wrinkled; sharp angles.	larger, darker flowers
		Other mallows	Malva spp	LEAF: sharply 5–7-lobed. FLOWER: **6–9** outer sepals.	number of outer sepals

CAMPANULACEAE	216	Venus's-looking-glass	Legousia hybrida	PLANT: stem with **short hairs**; 3–25 cm. FLOWER: 10 mm wide; sepals longer than petals. FRUIT: almost cylindrical; 30 mm long.	seed-shape diagnostic
	218	Greater Venus's-looking-glass	Legousia speculum-veneris	PLANT: stem **hairless**; 10–40 cm. FLOWER: 25 mm wide; sepals as long as petals. FRUIT: almost cylindrical; 30 mm long.	larger flower; shorter fruit
BORAGINACEAE	136	Field Gromwell	Lithospermum arvense	PLANT: usually erect, occasionally scrambling; bristly. LEAF: narrow, broader near the tip; **veins not apparent**. SEED: triangular and rounded; greyish black.	lack of leaf veins; seed shape
		Common Gromwell	Lithospermum officinale	PLANT: usually erect, occasionally scrambling; bristly. LEAF: narrow with **conspicuous veins**. SEED: oval; white.	leaf veins distinctive; habitat usually wood edges and hedges; seed shape
CONVOLVULACEAE		Hedge Bindweed	Calystegia sepium	LEAF: green; heart-shaped; 8–15 cm. FLOWER: white-pink, funnel-shaped, large; 3–4 cm wide.	FROM BLACK-BINDWEED (page 58): flower shape/size
		Field Bindweed	Convolvulus arvensis	LEAF: grey-green; arrowhead-shaped; 3–4 cm. FLOWER: white, funnel-shaped, large; 3–5 cm wide.	FROM BLACK-BINDWEED: flower shape/size

Flowers with **5 equal** petals: 3–30mm diameter — KEY DIFFERENCES

RANUNCULACEAE	62	Corn Buttercup	Ranunculus arvensis	PLANT: much-branched. LEAF: 3–5-lobed. FLOWER: bright lemon yellow; 12mm wide. SEED: round, covered in 2mm spines; 8mm long.	seed shape diagnostic (left)
	64	Hairy Buttercup	Ranunculus sardous	PLANT: erect, branched. LEAF: hairy, toothed lower 3-lobed; upper less lobed. FLOWER: 5-petalled; pale yellow; sepals bent back; 12–25mm wide. SEED: round, no spines; 2–3mm wide.	no stem tuber at base; seed shape (left)
		Bulbous Buttercup	Ranunculus bulbosus	PLANT: erect. LEAF: 3-lobed, middle lobe long stalked, rounded in outline. FLOWER: bright yellow; sepals bent back; 15–30mm wide. SEED: ovate with short beak; 3mm wide.	tuberous corm at base
	66	Prickly-fruited Buttercup	Ranunculus muricatus	PLANT: erect. LEAF: lower 3–5-lobed; upper less lobed. FLOWER: yellow; sepals bent back; 15mm wide. SEED: ovate with spines and a hooked beak at one end; 7–8mm long.	seed shape diagnostic (left)
	68	Small-flowered Buttercup	Ranunculus parviflorus	PLANT: spreading, branched. LEAF: yellowish-green, shallow lobes. FLOWER: pale yellow; hairy sepals bent back; 3mm wide. SEED: round with a short beak and covered in short hooks; 2·5–3·0mm.	small flowers and seed shape diagnostic (left)
		Creeping Buttercup	Ranunculus repens	PLANT: erect, stoloniferous. LEAF: 3-lobed, middle lobe long stalked, triangular in outline. FLOWER: glossy, golden yellow; hairy sepals not bent back; 20–30mm wide. SEED: roundish with short curved beak.	stolons; leaf shape, sepals not bent back
		Meadow Buttercup	Ranunculus acris	PLANT: erect, not stoloniferous. LEAF: 2–7-lobed, pentagonal or rounded in outline. FLOWER: glossy bright yellow; hairy sepals not bent back; 18–25mm wide. SEED: rounded with short, hooked beak.	leaf shape; sepals not bent back
	156	Mousetail	Myosurus minimus	PLANT: erect, from basal rosette; 2–15cm tall. LEAF: narrow, linear rosette leaves; flowering stem leafless. FLOWER: inconspicuous, small yellow-green petals which soon fall. SEED: narrow, conical seed-bearing receptacle elongates after flowering.	'mouse's-tail' seed receptacle is unmistakable (left)

Flowers with **5 unequal** petals and spur: 10-25mm diameter — KEY DIFFERENCES

VIOLACEAE	158	Field Pansy	Viola arvensis	FLOWER: 8–20mm; pale yellow/cream in colour with deep yellow or occasionally violet-blue centre; upper petals sometimes pale blue or purple; petals **inclined to form shallow cup**; sepals **as long as** petals.	FROM WILD PANSY: flower shape; sepal length
	160	Wild Pansy	Viola tricolor	FLOWER: 15–25mm; violet blue or tricoloured violet/pink, white and yellow; petals **flat**; sepals **shorter than** petals.	FROM FIELD PANSY: flower shape; sepal length
RANUNCULACEAE	150	Larkspur	Consolida ajacis	LEAF: finely-divided. FLOWER: 2–5mm; bracts **as long as** flower stalks. SEED POD: hairy.	bract length and seed pod characteristics
		Forking Larkspur	Consolida regalis	LEAF: 3-lobed. FLOWER: bracts **longer** than flower stalks. SEED POD: hairless.	bract length and seed pod characteristics
		Eastern Larkspur	Consolida orientalis	LEAF: 3-lobed. FLOWER: bracts much **shorter** than flower stalks. SEED POD: hairless.	bract length and seed pod characteristics

					KEY DIFFERENCES
LYTHRACEAE	134	Grass-poly	Lythrum hyssopifolia	PLANT: erect with branched stem, sometimes prostrate; in damp grassland. LEAF: small; lower leaves oval, upper leaves narrow and oval or spear-shaped; large leafy stipules at base of true leaves. FLOWER: 6-petalled; pink; usually 2–3 clustered; 5 mm wide.	can be confused with Knotgrass (**page 233**), which differs in having papery stipules which sheath the stem at the bases of the leaves and flowers 1–2 clustered
RANUNCULACEAE	168	Pheasant's-eye	Adonis annua	LEAF: bright green; finely dissected. FLOWER: 5–8-petalled; deep red with dark basal spots on petals; resembles an anemone; 15–25 mm wide. SEED-HEAD: elongated oval.	unmistakable when flowering; when not in flower could be mistaken for a mayweed (**page 242**) but these are hairy, have broader leaf segments and are a much darker green in colour

					KEY DIFFERENCES
APIACEAE	60	Small Bur-parsley	Caucalis platycarpos	LEAF: lower leaves oval, upper leaves narrow and parallel-sided. FLOWER: white. UMBEL: 2–5 rays; 0–2 bracts. SEED: with long curved spines, formed in pairs; 13 mm long.	seed shape diagnostic
	140	Knotted Hedge-parsley	Torilis nodosa	PLANT: erect; stem solid; up to 30 cm. LEAF: pinnately divided. FLOWER: pinkish-white in dense **stalkless** clusters along the stem; 1 mm. UMBEL: 5–10 mm. SEED: oval with long or short spines; 3 mm wide.	flower clusters characteristic
	142	Spreading Hedge-parsley	Torilis arvensis	PLANT: spreading. branched; stem solid; up to 50 cm. LEAF: 2-pinnate in basal rosette and on stem. FLOWER: white or pinkish with unequal petals; 2 mm. UMBEL: 3–5 rays; **0–1 bracts**; 10–25 mm. SEED: oval with long hooked spines; 4–6 mm long.	fruit characteristic; few bracts
		Upright Hedge-parsley	Torilis japonica	PLANT: erect, much-branched; stem solid; covered in short bristles; up to 50 cm. LEAF: 2-pinnate. FLOWER: pinkish-white; outer petals larger than inner; 2–3 mm. UMBEL: 5–12 rays; **4–6 bracts**. SEED: oval with hooked spines, formed in pairs; 3–4 mm wide.	leaves larger; more bracts at base of umbel
		Fool's Parsley	Aethusa cynapium	PLANT: erect; stem solid; up to 125 cm. LEAF: 1–3-pinnate. FLOWER: pinkish-white with unequal petals; 2 mm. UMBEL: 4–20 rays; 0 bracts; **long bracteoles**; 20–60 mm. SEED: seeds with long curved spines, formed in pairs; 13 mm long.	distinctive bracteoles give flowering plants a bearded effect
		Wild Carrot	Daucus carota	PLANT: erect; stem hollow; up to 120 cm. LEAF: 3-pinnate. FLOWER: white; central umbel flower often pink; 2 mm. UMBEL: numerous rays; **7–13 pinnate bracts**; 30–70mm.	distinctive red-purple flower in centre of flowerhead; pinnate bracts
		Cow Parsley	Anthriscus sylvestris	PLANT: erect; stem solid; up to 30–100 cm. LEAF: 2–3-pinnate FLOWER: white; 3–4 mm. UMBEL: 4–10 rays; **0 bracts**; several bracteoles; 20–60mm.	absence of bracts and presence of bracteoles

Flowers tiny with 5 ± equal petals: IN UMBRELLA-LIKE CLUSTERS — KEY DIFFERENCES

APIACEAE					
	162	Corn Parsley	Petroselinum segetum	PLANT: erect; stem **hollow**; hairy below, hairless above; green to purple; up to 100 cm. LEAF: pinnately divided; **smell of parsley** if crushed. FLOWER: greenish-white; 1 mm. UMBEL: 3–6 rays; 2–5 bracts; 2–5 bracteoles; 10–50 mm.	distinctive scent of parsley from crushed leaves; highly irregular umbels
		Stone Parsley	Sison amomum	PLANT: erect; stem **solid**; hairless; greyish-green; up to 80 cm. LEAF: pinnately divided; **smell of petrol** if crushed. FLOWER: white; 1 mm. UMBEL: 3–6 rays; 2–4 bracts; 2–4 bracteoles; 10–40 mm.	distinctive unpleasant scent of petrol from crushed leaves
	170	Greater Pignut	Bunium bulbocastaneum	PLANT: erect; much-branched; **stem solid**; hairless; 50–100 cm. LEAF: 3-pinnate; **very finely divided**; sheathing stem. FLOWER: white; 2 mm. UMBEL: 10–20 rays; numerous bracts and bracteoles; 30–80 mm.	characteristic leaf shape
		Common Pignut	Conopodium majus	PLANT: erect; much-branched; **stem hollow after flowering**; hairless; up to 60 cm. LEAF: 2-pinnate; **very finely divided**; sheathing stem. FLOWER: white; 1–3 mm. UMBEL: 6–12 rays; 0–5 bracts and bracteoles. SEED: seeds with long curved spines, formed in pairs; 13 mm long; 30–70 mm.	FROM GREATER PIGNUT Hollow stem, fewer bracts and bracteoles
	184	Shepherd's-needle	Scandix pecten-veneris	LEAF: finely divided; narrow, parallel segments; leaves join round stem. FLOWER: white; 1 mm. UMBEL: 2 rays; 0 bracts; numerous bracteoles. SEED: characteristic, elongated needle-like fruit; up to 50 mm long.	seed shape diagnostic
	210	Thorow-wax	Bupleurum rotundifolium	LEAF: **round to oval**; upper leaves surround stem. FLOWER: yellowish; clusters surrounded by green petal-like bracteoles.	FROM FALSE THOROW-WAX: leaf shape
		False Thorow-wax	Bupleurum subovatum	LEAF: **narrowly oval**. FLOWER: yellowish; clusters surrounded by green petal-like bracteoles.	leaf shape

Flowers with petals ± fused: BELL-SHAPED COROLLA — KEY DIFFERENCES

BORAGINACEAE					
	222	Purple Viper's-bugloss	Echium plantagineum	LEAF: base leaves oval with distinct lateral veins; stem leaves narrowly oval; up to 15 cm long by 3 cm wide. FLOWER: bell-shaped; 2 projecting stamens; 20–25 mm long. SEED: seed pyramidal; pitted and warty.	FROM VIPER'S-BUGLOSS: number of stamens
		Viper's-bugloss	Echium vulgare	LEAF: base leaves oval with no apparent lateral veins; stem leaves narrowly oval; 40 mm long by 4 mm wide. FLOWER: bell-shaped; 3–5 vivid pink projecting stamens; 15–18 mm long. SEED: angular; wrinkled.	FROM PURPLE VIPER'S-BUGLOSS: number of stamens

Although fumitory species (Fumariaceae) are well-defined they are sometimes a difficult group to identify due to the differences between species being subtle across a range of features. Individual plants within a species can show a high degree of variability in these features depending on many factors. These factors include for example, how shaded the area is in which a plant grows since this can affect the number of flowers produced and the degree of recurvedness in the fruit stalks.

The following key does not cover all these variations, but sets out the features that should be looked at when making an identification. The illustrations are based on photographs and detailed specimen drawings. The flowers are are shown at 2× life-size and the fruits at 3× life-size.

A full description of the species can be found on the relevant page to which the entries in the key are cross-referenced.

Fumitory flower and leaves showing key features as described in the following text.

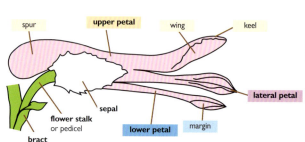

spur · upper petal · wing · keel · sepal · lateral petal · flower stalk or pedicel · lower petal · margin · bract

very narrow; channelled leaf segment

broad; unfolded leaf segment

	Common Fumitory *Fumaria officinalis*	Dense-flowered Fumitory *Fumaria densiflora*	Few-flowered Fumitory *Fumaria vaillantii*	Fine-leaved Fumitory *Fumaria parviflora*
Page No.	114	116	118	120
LEAF SEGMENTS	channelled	channelled; very narrow	relatively flat	channelled; very narrow
FLOWER COLOUR	pinkish-red with dark red tipped petals	pink/red with black-tipped petals	**pale pink** with reddish-black tipped petals.	**white** with reddish-black-tipped petals
FLOWER LENGTH	7—8mm	6—7mm	5—6mm	5—6mm
FLOWER ×2				
SEPAL	toothed; white or pink	large; white	toothed; pale purple	toothed; white
SEPAL SIZE	1·0—1·5mm × 1·5—3·5mm	**2 × 3mm**	0·5mm × 1·0mm	0·8mm × 1·0mm
FLOWERS PER RACEME	**10—40, usually more than 20**	20—25	**6—16**, sometimes more	**16—20**
FRUIT	distinctly wider than long with flattened or slightly notched apex	fruit spherical; rounded at apex	fruit spherical; rounded at apex	fruit spherical; sometimes with short beak at apex
SIZE OF FRUIT	2·5mm diameter	2mm diameter	2mm diameter	2mm diameter
FRUIT ×3				
KEY DIFFERENCES	small flower with small sepals; number of flowers in raceme	small flower with large sepals	small flower with small sepals; number of flowers in raceme	small flower; flower colour

	Common Ramping-fumitory Fumaria muralis ssp. boraei	Martin's Ramping-fumitory Fumaria reuteri	Purple Ramping-fumitory Fumaria purpurea	Tall Ramping-fumitory Fumaria bastardii
Page No.	**122**	**124**	**126**	**128**
LEAF SEGMENTS	relatively broad and flat	relatively broad and flat	relatively broad and flat	relatively broad and flat
FLOWER COLOUR	pinkish-red with darker red tipped petals	pink with blackish-red tipped petals	pink-purple with darker purple tipped petals.	salmon-pink with **dark tips to lateral petals only**
FLOWER LENGTH	9–11 mm	11–13 mm	10–13 mm	9–11 mm
FLOWER ×2				
SEPAL	toothed mainly at the base; pale	**very few teeth**; white	toothed; oblong; white	toothed; white
SEPAL SIZE	1·5–3·0 mm × 3·0–5·0 mm	2·5 mm × 4·0 mm	6 mm × 3 mm	2 mm × 3 mm
FLOWERS PER RACEME	**12–15**; raceme **shorter** than stalk	15–20; raceme **longer** than stalk	20–25; raceme same length as stalk	10–18; raceme longer than stalk
FRUIT	fruit spherical with rounded apex	fruit spherical	fruit spherical on recurved fruit stalk	fruit spherical fruit stalk **never** recurved
SIZE OF FRUIT	2·5 mm diameter	2·5 mm diameter	2·5 mm diameter	2·5 mm diameter
FRUIT ×3				
KEY DIFFERENCES	raceme shorter than stalk; number of flowers per raceme;	sepal with few teeth + raceme longer than stalk	frequently recurved seed stalk; large sepal	small flower + small sepal + raceme longer than stalk

	Western Ramping-fumitory Fumaria occidentalis	White Ramping-fumitory Fumaria capreolata
Page No.	**130**	
LEAF SEGMENTS	relatively broad and flat	relatively broad and flat
FLOWER COLOUR	white, becoming pink with blackish-red tipped lateral petals	creamy-white with blackish-red tipped petals
FLOWER LENGTH	12–15 mm	10–14 mm
FLOWER ×2		
SEPAL	toothed at base; white	toothed at base; white
SEPAL SIZE	5 mm × 3 mm	4–6 mm × 2·5–3·0 mm
FLOWERS PER RACEME	12–20; raceme same length as stalk	ca. 20; raceme shorter than stalk
FRUIT	fruit spherical with slight projection at apex	fruit spherical on **strongly recurved stalk**
SIZE OF FRUIT	3 mm diameter	2 mm diameter
FRUIT ×3	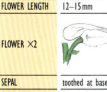	
KEY DIFFERENCES	FROM F. CAPREOLATA: less flowers per raceme; flower colour never creamy-white; larger flower with smaller sepal	FROM F. OCCIDENTALIS: creamy-white flower colour distinctive; raceme shorter than stalk; recurved fruit stalk

White Ramping-fumitory Fumaria capreolata
This species is scattered throughout Britain but is predominantly coastal and most likely to be encountered in Northern Ireland, Wales and south-west England. Unlike other fumitories it can be a winter annual.

239

FABACEAE	208	Slender Tare	Vicia parviflora	PLANT: scrambling; up to 60 cm. LEAF: 25mm long; divided into 2–5 pairs of leaflets. FLOWER: **pale purple**; groups of 2–8; 6–8mm long; stalks longer than leaves. SEED POD: **5–8 seeds**; 15–18mm long.	FROM HAIRY TARE AND SMOOTH TARE: flower colour and seed pod characteristics (left)
		Hairy Tare	Vicia hirsuta	PLANT: scrambling; up to 70 cm. LEAF: 5–12mm long; divided into 4–8 pairs of leaflets. FLOWER: **white-purplish**; groups of 1–9; 4–5mm long; stalks longer than leaves. SEED POD: **2 seeds**; 10mm long.	flower colour and seed pod characteristics (left)
		Smooth Tare	Vicia tetrasperma	PLANT: scrambling; up to 60 cm. LEAF: 10–20mm long; divided into 3–6 pairs of leaflets FLOWER: **very pale blue**; groups of 1–2; 4mm long; stalks longer than leaves SEED POD: **4 seeds**; 12–15mm long.	flower colour and seed pod characteristics (left)
	220	Yellow Vetchling	Lathyrus aphaca	PLANT: scrambling; up to 100 cm. LEAF: leaves absent in mature plants; large, **triangular leaf-like stipules** present in pairs along the stem; stipules 10–30mm long FLOWER: yellow; solitary; 10–12mm long. SEED POD: 6–8 seeds; 20–30mm long.	FROM MEADOW VETCHLING: single flower; leaf-like stipules
		Meadow Vetchling	Lathyrus pratensis	PLANT: scrambling; up to 120 cm. LEAF: narrow, spear-shaped leaves; leaf-like **stipules arrowhead-shaped**; leaflets 10–30mm long; stipules 10–25mm long. FLOWER: yellow; clusters of 5–12 flowers; 15–18mm long. SEED POD: 5–10 seeds; 25–35mm long.	cluster of flowers; true leaves and distinctive stipule shape

SCROPHULARIACEAE	110	Round-leaved Fluellen	Kickxia spuria	LEAF: **oval**, hairy; arranged alternately. FLOWER: yellow with a deep purple upper lip, spur **straight**; 8–11mm long.	FROM ROUND-LEAVED FLUELLEN: leaf shape.
	112	Sharp-leaved Fluellen	Kickxia elatine	LEAF: **arrowhead-shaped**, hairy; arranged alternately. FLOWER: yellow with a purple upper lip, spur **curved**; 7–9mm long.	FROM SHARP-LEAVED FLUELLEN: leaf shape.
	212	Small Toadflax	Chaenorhinum minus	FLOWER: purple outside, paler inside; solitary; short spur; corolla 6–9mm.	distinctive flower coloration
	224	Weasel's-snout OR Lesser Snapdragon	Misopates orontium	FLOWER: deep pink; snapdragon shape; stalkless; 15mm across.	flower distinctive
	94	Field Cow-wheat	Melampyrum arvense	PLANT: up to 60 cm. LEAF: spear-shaped; stalkless; glossy-green. FLOWER: yellow corolla-tube; purple-pink lips; bracts pinkish-red with very rough-toothed outline; cylindrical flower spike; corolla 20–24mm.	FROM CRESTED COW-WHEAT: Shape of flower spike and bracts
		Crested Cow-wheat	Malampyrum cristatum	PLANT: 20–50 cm. LEAF: 3-lobed, middle lobe long stalked, rounded in outline. FLOWER: yellow corolla-tube; purple-pink lips; bracts purple with fine-toothed outline; four-sided flower spike; corolla 12–16mm.	FROM FIELD COW-WHEAT: Shape of flower spike and bracts; habitat wood margins

Flowers with petals ± **fused**: LAMIACEAE; Labiate family				KEY DIFFERENCES
LAMIACEAE	**104** Cut-leaved Dead-nettle	Lamium hybridum	PLANT: branched irregularly; to 30 cm. LEAF: triangular with deep, irregular teeth; short stalks. FLOWER: pinkish-purple; 8—12 mm long.	FROM OTHER DEAD-NETTLES: leaf shape
	Red Dead-nettle	Lamium purpureum	PLANT: branched from the base; to 45 cm. LEAF: rounded with shallow, even teeth; short stalks. FLOWER: purple; calyx shorter than corolla tube; 5—6 mm long.	leaf shape; calyx shorter than corolla tube
	106 Henbit Dead-nettle	Lamium amplexicaule	PLANT: erect, branched irregularly; to 30 cm. LEAF: rounded with shallow, irregular teeth; lower leaves with short stalks, upper leaves stalkless, **clasping the stem**. FLOWER: pinkish-purple; hairy calyx; 14 mm long.	FROM OTHER DEAD-NETTLES: leaves which clasp the stem diagnostic
	108 Northern Dead-nettle	Lamium confertum	PLANT: branched irregularly; to 30 cm. LEAF: triangular in shape with deep, irregular teeth; short stalks. FLOWER: purple; calyx as long as corolla tube; **10—18 mm long**.	FROM OTHER DEAD-NETTLES: large flowers; calyx as long as corolla tube
	132 Cut-leaved Germander	Teucrium botrys	PLANT: much-branched, hairy; to 30 cm. LEAF: oval in shape, **deeply divided**. FLOWER: in whorls; pinkish-red; corolla tube enclosed in calyx; 6 mm long.	unmistakable
	226 Field Woundwort	Stachys arvensis	PLANT: weak spreading stems; branched from base; up to 20 cm. LEAF: oval to heart-shaped; irregularly toothed edges; hairy; short stalks; arranged as opposite pairs. FLOWER: pale purple; in whorls of 2—6; 6—7 mm long.	easily confused with other members of the mint family; although this species has smaller flowers than other pale-flowered species; Corn Mint Mentha arvensis is perhaps the most similar species in the arable habitat but differs, like other Mentha mints, in having radially symmetrical flowers and a light minty smell
	Common Hemp-nettle	Galeopsis tetrahit	PLANT: stems thickened at leaf junctions; 10—100 cm. LEAF: upper narrowly oval; serrated edge. FLOWER: pink, purple or white; corolla slightly longer than calyx; 13—20 mm long.	flower colour and relatively short corolla tube distinctive
	146 Red Hemp-nettle	Galeopsis angustifolia	PLANT: not swollen at leaf nodes; to 50 cm. LEAF: narrowly spear-shaped with 1—4 teeth on each edge; lightly hairy; short stalk. FLOWER: pinkish-red; 15—25 mm long.	flower colour and length distinctive; when not in flower Red Bartsia Odontites verna is superficially similar but has leaves without stalks and a single-sided flower spike
	144 Downy Hemp-nettle	Galeopsis segetum	PLANT: up to 50 cm. LEAF: narrow, oval-shaped with 3—9 teeth on each edge; velvety beneath; short stalk. FLOWER: sulphur yellow; long corolla tube, 4× calyx length; large lower lip; 20—30 mm long.	flower colour; FROM LARGE-FLOWERED HEMP-NETTLE: corolla/calyx length ratio distinctive
	Large-flowered Hemp-nettle	Galeopsis speciosa	PLANT: stem thickened below nodes; 10—100 cm. LEAF: upper narrowly oval. FLOWER: pale yellow with purple lower lip; long corolla tube, 2× calyx length; large lower lip; 20—40 mm long.	flower colour; FROM DOWNY HEMP-NETTLE: corolla/calyx length ratio distinctive
	138 Ground-pine	Ajuga chamaepitys	PLANT: up to 20 cm. LEAF: 3 linear lobes. FLOWER: yellow upper lip and yellow, spotted red, lower lip.	unmistakable, resembles a pine seedling in look and smell

ASTERACEAE					
	72 Smooth Cat's-ear	Hypochaeris glabra	PLANT: Dandelion like, erect stems growing from a basal rosette. LEAF: spear-shaped, narrow, **not hairy**. FLOWER-HEAD: yellow; individual floret petal **2×** as long as broad; 5—10mm across.	FROM OTHER 'DANDELIONS': basal leaf rosette plus bract-like scales on stem; FROM COMMON CAT'S-EAR: smooth leaves; smaller flower-head
	Common Cat's-ear	Hypochaeris radicata	PLANT: Dandelion like, erect stems growing from a basal rosette. LEAF: spear-shaped, narrow, **hairy**. FLOWER-HEAD: yellow; individual floret petal **2—4×** as long as broad; 25—40mm across.	FROM SMOOTH CAT'S-EAR: hairy leaves; larger flower-head
	206 Lamb's Succory	Arnoseris minima	PLANT: erect; stems from basal rosette; stems **hollow and inflated** below flower-head. LEAF: oblong with few teeth; broader near the tip. FLOWER-HEAD: 7—11mm across.	superficially resembles other dandelion-type flowers; however, the inflated stem is diagnostic
	154 Corn Marigold	Anthemis arvensis	PLANT: erect, much-branched; hairless with waxy surface; blue-green appearance. FLOWER-HEAD: disc florets and ray florets golden-yellow; 30—60mm across.	unmistakable
	78 Corn Chamomile	Chrysanthemum segetum	PLANT: much branched, softly hairy. LEAF: finely divided; 1—3-pinnate. FLOWER-HEAD: yellow central florets and white ray florets; **broad chaffy scales** in yellow florets; 20—40mm across.	leaves with **pleasant chamomile-like scent**; broad chaffy scales present (left)
	Stinking Mayweed	Anthemis cotula	PLANT: branched; sparsely hairy. LEAF: finely divided; 1—3-pinnate. FLOWER-HEAD: yellow central florets and white ray florets; **narrow chaffy scales** in yellow florets; 12—25mm across.	leaves with **unpleasant smell when crushed**; narrow chaffy scales present (left)
	Scentless Mayweed	Tripleurospermum inodurum	PLANT: simply-branched; hairless. LEAF: finely divided; 2—3-pinnate. FLOWER-HEAD: yellow central florets and white ray florets; **no chaffy scales** in yellow florets; 12—25mm across.	leaves with **no scent when crushed**; no chaffy scales
	Scented Mayweed	Matricaria recutita	PLANT: much-branched; hairless. LEAF: finely divided; 2—3-pinnate. FLOWER-HEAD: yellow central florets and white ray florets; **broad chaffy scales** in yellow florets; 12—22mm across.	leaves with **pleasant chamomile-like scent;** no chaffy scales present
	84 Cornflower	Centaurea cyanus	PLANT: erect, much-branched; hairy; grey-geen LEAF: long, narrow, variably dissected. FLOWER-HEAD: **divided** outline; intense blue outer florets; purple inner florets; 30mm across.	superficially similar to Field Scabious but unlikely to be mistaken
	Field Scabious	Knautia arvensis	PLANT: erect. LEAF: pinnately-lobed FLOWER-HEAD: **roundish** outline; bluish-lilac, outer florets larger than inner florets; 40mm across.	flowers never intense blue; leaf shape

Identification key to arable plants

ASTERACEAE					
	98	Broad-leaved Cudweed	Filago pyramidata	PLANT: branched from base; covered in a felt of grey hairs; up to 30 cm. LEAF: **narrow, broader near the tip**; up to 20 mm long by 3—4 mm wide. FLOWERS: tiny, 2—7 clustered in dense, grey woolly flower-heads with narrow pointed yellow-brown bracts.	leaf shape (left)
	100	Narrow-leaved Cudweed	Filago gallica	PLANT: branched from base; covered in fine grey hairs; up to 25 cm. LEAF: **very narrow;** up to 18 mm long by 1 mm wide. FLOWERS: tiny, 2—6 clustered in dense, grey, woolly flowerheads.	leaf shape (left)
	102	Red-tipped Cudweed	Filago lutescens	PLANT: branched; covered in a felt of yellow-tinged grey hairs; up to 25 cm. LEAF: **parallel-sided**; up to 20 mm long by 3—4 mm wide. FLOWERS: tiny, 10—20 clustered in dense, grey, woolly flower-heads; narrow pointed **bracts with red tips**.	leaf shape (left); red-tipped bracts
		Common Cudweed	Filago vulgaris	PLANT: branched; covered in silvery-white hairs; up to 30 cm. LEAF: **wavy-edged**, narrow, tapering to a point; up to 15 mm long. FLOWERS: tiny, 8—15 clustered in dense, **white** woolly flower-heads; narrow pointed bracts with yellowish tips.	leaf shape (left); whitish hairs
		Small Cudweed	Filago minima	PLANT: branched; covered in silky grey hairs; up to 30 cm. LEAF: narrow; up to **10 mm** long. FLOWERS: tiny, 3—6 clustered in dense flower-heads; narrow blunt bracts with yellowish tips.	leaf length short

Field Brome

Bromus arvensis

IDENTIFICATION:

All species of brome grass are similar in overall structure. Field Brome is usually a robust grass growing up to 1 m tall. The leaves are up to 20 cm long and 5 mm wide, with a blunt ligule up to 4 mm long. Both leaf and stem are softly and densely hairy. The inflorescences are much-branched with drooping, long-stalked spikelets. The spikelets are oval and slightly laterally compressed, measuring approximately 1·5 cm long by 4 mm wide, with anthers up to 5 mm long. The lemma has inrolled edges, and is less than 6·5 mm long; it has a rough, straight awn up to 1 cm long.

Similar species: Brome grasses are a very difficult group to identify. Barren Brome and Great Brome have narrower, parallel-sided spikelets. The very common Soft-brome and Meadow Brome have much smaller spikelets. The most similar species, Rye Brome, has lemmas up to 9 mm long and short anthers.

HABITAT:

Arable field margins.

SOIL TYPE:

Loams and clay loams.

DISTRIBUTION:
Poorly known.

LIFE CYCLE:
Flowers from June to July.
In common with other bromes, seed longevity probably poor.
Germination is mainly in the autumn.

J F M A M J J A S O N D

REASONS FOR DECLINE:
This species may have relied on harvest and resowing along with crop seed, and it would have started to decline with the introduction of efficient seed-cleaning machinery in the late 19th century. It may also be susceptible to selective grass-specific herbicides.

Field Brome: lemma *(left)* ×2, spikelet *(right)* ×2

Great Brome

Anisantha diandra

IDENTIFICATION:

All species of brome grass are similar in overall structure. Great Brome is a robust grass growing up to 80 cm tall. The leaves are up to 20 cm long and 4–8 mm wide, with a blunt ligule up to 6 mm long. Both leaf and stem are softly and densely hairy. The inflorescences are much-branched with erect, long-stalked spikelets. The spikelets are rectangular and laterally compressed, measuring approximately 3–4 cm long, with a rough, straight awn up to 5 cm long.

Similar species: Brome grasses are a very difficult group to identify. The most similar species is the very common Barren Brome, which has spikelets less than 3 cm long, with awns up to 2·5 cm long. Other species of brome have oval spikelets.

HABITAT:

Arable field margins, gardens and tracksides.

SOIL TYPE:

Sands and sandy loams.

DISTRIBUTION:

This species is most widespread in East Anglia. It also occurs in scattered sites northwards to Yorkshire, and is now known to be more widespread throughout southern Britain.

Great Brome: lemma (*left*) ×2, spikelet (*right*) × 1·3

LIFE CYCLE:

Flowers from May to June. Seed is likely to be short-lived. Germination is probably entirely in the autumn.

J F M A M J J A S O N D

REASONS FOR DECLINE:

Great Brome is thought to have actually increased in recent years.

246

Interrupted Brome

Bromus interruptus

IDENTIFICATION:

All species of brome grass are similar in overall structure. Interrupted Brome is usually a robust grass growing up to 1m tall, but is sometimes much smaller. The leaves are up to 20 cm long and 5 mm wide, with a blunt ligule up to 4 mm long. Both leaf and stem are softly and densely hairy. The inflorescences are branched with very short-stalked spikelets in clusters of three. The spikelets are oval and slightly laterally compressed, measuring approximately 1·0–1·7 cm long. The lemma has a rough, straight awn up to 8 mm in length.

Similar species: It resembles other brome grasses, particularly the common Soft-brome, but the inflorescences are very distinctive.

HABITAT:

Formerly in arable fields.

SOIL TYPE:

Various, usually clay loams.

E?

DISTRIBUTION:

Interrupted Brome was formerly found in scattered sites through central-southern England and East Anglia. It was an endemic species but is now thought to be extinct. It is maintained in cultivation at the Royal Botanic Gardens, Kew.

LIFE CYCLE:

Flowers from June to July. Seed is very short-lived. Germination is in the autumn.

J F M A M J J A S O N D

REASONS FOR DECLINE:

In comparison to other bromes, Interrupted Brome is poorly competitive, and will have been affected by the development of highly competitive crops and the increasing application of nitrogen.

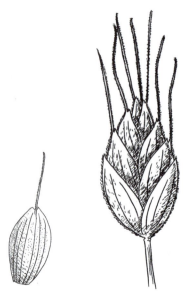

Interrupted Brome: lemma (*left*) × 2, spikelet (*right*) × 2

Rye Brome

Bromus secalinus

IDENTIFICATION:

All species of brome grass are similar in overall structure. Rye Brome is usually a robust grass growing up to 1m tall. The leaves are up to 20 cm long and 5 mm wide, with a blunt ligule up to 4 mm long. Both leaf and stem are softly and densely hairy. The inflorescences are much-branched with long stalked spikelets. The spikelets are oval and slightly laterally compressed, measuring approximately 1·5 cm long by 4 mm wide, with anthers up to 1 mm long. The lemma has inrolled edges, and is up to 9 mm long. It has a rough, straight awn up to 1 cm long.

Similar species: Brome grasses are a very difficult group to identify. Barren Brome and Great Brome have narrower, parallel-sided spikelets. The very common Soft-brome and Meadow Brome have much smaller spikelets. The most similar species, Field Brome, has lemmas less than 6·5 mm long and long anthers.

HABITAT:

Arable fields.

SOIL TYPE:

Various, but mainly heavy clay loams.

DISTRIBUTION:

Poorly known, but scattered throughout southern Britain. It can occur in large patches, and is more abundant than the closely related Field Brome.

LIFE CYCLE:

Flowers from June to July.
Seed is thought to be short-lived.
Germination is in the autumn.

J F M A M J J A S O N D

REASONS FOR DECLINE:

Rye Brome is thought to have been harvested and resown with the crop seed. Its decline can probably be attributed largely to improved seed cleaning technology introduced at the end of the 19th century.

Rye Brome: lemma (*left*) ×2, spikelet (*right*) ×2

Darnel

Lolium temulentum

E?

IDENTIFICATION:
An erect grass, with solitary or tufted stems. The leaves are up to 40 cm in length, green, and hairless. The ligules, at the junction of the leaf and stem, are 2 mm in length and blunt, There is also a pair of membraneous auricles clasping the stem. The flowers are stalkless and borne alternately in a stiff spike up to 30 cm long. The individual spikelets are 12–20 mm long and 4–6 mm wide. Each spikelet has an enclosing glume that exceeds it in length. The lemmas usually have straight, rough awns up to 2 cm long, but are occasionally awnless.

Similar species: Perennial and Italian Rye-grasses. Perennial Rye-grass flowers are not awned. In addition, the length of Italian Rye-grass the glumes do not exceed the spikelet.

HABITAT:
Cereal fields in the past. Now occasional on rubbish tips.

SOIL TYPE:
Various.

DISTRIBUTION:
Now extinct in Britain except as a casual. Formerly scattered throughout the country, especially in the south. Still found in the Irish Aran Islands.

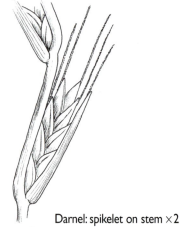

Darnel: spikelet on stem ×2

LIFE CYCLE:
Flowers from June to August. Seed is probably short-lived. Germination period is not known.

J	F	M	A	M	J	J	A	S	O	N	D

REASONS FOR DECLINE:
In the past, Darnel may have been harvested and resown with crop seed. Much of its decline may have been due to the introduction of efficient seed-cleaning machinery.

Nit-grass

Gastridium ventricosum

IDENTIFICATION:

Nit-grass is an erect grass up to 50 cm tall which usually grows in tufts. The leaves are up to 10 cm long and 4 mm wide, but are usually smaller, with a short ligule. The flowers are borne in a compact terminal spike which is spear-shaped and tapers to a point. The lemmas usually have a bent awn up to 4 mm in length.

Similar species: Other grasses of arable land with flowers in compact spikes are the very common Black-grass which has long, cylindrical spikes, and Annual Vernal-grass which has distinctive tufts of hairs at the junction of leaf and stem.

Associated uncommon species: Lesser Quaking-grass also occurs at its only known site in Hampshire.

HABITAT:

Arable field margins, set-aside, drought-prone calcareous grassland especially near the coast.

SOIL TYPE:

Thin soils over (usually) Jurassic or Carboniferous limestones; calcareous clays where these are well-drained or prone to summer drought; and sometimes other soil types.

MANAGEMENT REQUIREMENTS:

Autumn cultivation.

Nit-grass: spikelet × 10

DISTRIBUTION:

As an arable species, Nit-grass was formerly widespread in the south and east of England. It now occurs in one known arable site in Hampshire. It has also recently occurred in set-aside land in Dorset and Somerset.

LIFE CYCLE:

Flowers from May to September. Seed longevity is unknown.
Germination is mainly in the autumn, although seedlings can be killed by hard frosts.

J F M A M J J A S O N D

REASONS FOR DECLINE:

Nit-grass has declined in arable habitats, probably as a result of increases in the amounts of nitrogen used and the development of competitive crop varieties. It may also be susceptible to grass-weed herbicides. However, it has increased in recent years in set-aside fields.

Lesser Quaking-grass

Briza minor

IDENTIFICATION:
An erect grass growing up to 50 cm tall. The leaves are 3–7 mm wide, hairless and blue-green in colour, with a pointed ligule 3–6 mm long. The much-branched inflorescence resembles that of the more common Quaking-grass (a plant of calcareous and neutral grasslands). The large triangular flower spikelets are borne at the end of the long inflorescence branches.

Similar species: This species is unmistakable in an arable context, but could be confused with the common Quaking-grass of grasslands.

Associated uncommon species: Usually in species-rich communities with Small-flowered Catchfly, Four-leaved Allseed (on the Isles of Scilly), Weasel's-snout and Corn Marigold. With Nit-grass at one arable site in Hampshire.

HABITAT:
Arable field margins. On the Isles of Scilly, in bulb and potato fields. The majority of sites are coastal.

SOIL TYPE:
Acidic sands and sandy loams and gravels.

DISTRIBUTION:
Lesser Quaking-grass has never been very widespread in Britain. It is now largely restricted to Cornwall, the Isles of Scilly and the Hampshire Basin, with isolated sites in Surrey and Kent.

LIFE CYCLE:
Flowers from June to August.
Seed longevity is unknown.
Germination is mainly in the autumn, although there is also some in the spring.

J F M A M J J A S O N D

REASONS FOR DECLINE:
This species has probably declined with the introduction of highly competitive crop varieties and the increase in application of nitrogen. The marginal arable sites that Lesser Quaking-grass occupies have also been susceptible to abandonment and conversion to grassland.

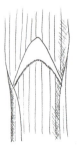

Lesser Quaking-grass: spikelet (*left*) ×2, ligule (*right*) ×5

Dense Silky-bent

Apera interrupta

IDENTIFICATION:

Dense Silky-bent is an erect annual grass, growing up to 70 cm tall in arable land although usually much shorter. The stems and leaves are largely hairless. The leaves can be up to 10 cm long and are up to 4 mm wide, with a long pointed ligule up to 5 mm long. The leaf-sheaths are green or purple. The inflorescence consists of a loose, interrupted spike with stalkless branches. Each spikelet has a very long, straight awn.

Similar species: Loose Silky-bent is similar, but has a much-branched inflorescence with long-stalked branches.

Associated uncommon species: In East Anglia, Dense Silky-bent occurs with Night-flowering Catchfly and Corn Chamomile. In Oxfordshire it occurs at one site with Prickly Poppy, Shepherd's-needle, Corn Marigold and Cornflower.

HABITAT:

It is most frequent in arable field margins but also occurs in a few places on railway tracks, tracks across heathland, coastal shingle and gravel pits.

SOIL TYPE:

Calcareous, sandy loams.

DISTRIBUTION:

Dense Silky-bent has never been a widespread or abundant species. It is locally frequent in Norfolk and Suffolk, particularly in Breckland, with a few sites in Oxfordshire, Lincolnshire and North Yorkshire.

LIFE CYCLE:

Flowers from June to July. Seed longevity is unknown. Germination period is in the autumn.

J F M A M J J A S O N D

REASONS FOR DECLINE:

This species may have increased a little in very recent years, although this may be simply a result of increased survey effort. Declines on arable land in the past have been due to the effects of grass-specific herbicides and highly competitive, fertilised crops.

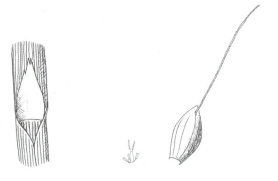

Dense Silky-bent:
ligule (*left*) ×3, anther (*centre*) ×10, awn (*right*) ×4

Loose Silky-bent
Apera spica-venti

IDENTIFICATION:
Loose Silky-bent is an erect grass growing up to 1m tall, with either solitary or multiple stems. The leaves and stems are hairless. The leaves can be up to 20 cm long by 10 mm wide and are often rough to the touch. The ligule is up to 1 cm long and the leaf-sheaths are green or purple. The inflorescence is very finely branched and widely spreading, up to 20 cm long and 15 cm wide. Each spikelet has a long, straight awn.

Similar species: Other arable grasses that have widely branched, spreading inflorescences with small spikelets are Black Bent and Creeping Bent. However, neither of these species have long-awned spikelets.

Associated uncommon species: Loose Silky-bent can be found alongside Dense Silky-bent and other uncommon species such as Prickly Poppy. In Breckland it occurs at sites with several Breckland rarities.

HABITAT:
Arable fields, including set-aside land, also on waste ground, tracks and roadsides.

SOIL TYPE:
Sandy loams and sands.

DISTRIBUTION:
Loose Silky-bent is still frequent on sandy soils around London, in East Anglia, Lincolnshire and South Yorkshire. It can be abundant in some localities, and may have increased in recent years.

LIFE CYCLE:
Flowers from May to September. Seed longevity is unknown. Germination is both in the autumn and spring.

J F M A M J J A S O N D

REASONS FOR DECLINE:
This species may have increased a little in very recent years, although this may be simply a result of increased survey effort. Declines on arable land in the past have been due to the effects of grass-specific herbicides.

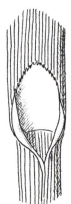

Loose Silky-bent:
anther (*right*) × 10 and ligule (*left*) × 3

Annual Vernal-grass
Anthoxanthum aristatum

IDENTIFICATION:
Annual Vernal-grass grows 10–40 cm tall with solitary, but sometimes multiple, stems. The leaves are 2–4 mm wide and up to 5 cm long. The leaves and the stem are usually hairless, although there are tufts of long hairs where the two meet. The spikelets have long and conspicuous bent awns and are gathered into a pale green terminal spike between 1 cm and 3 cm long. The spikelets

Similar species: The only other arable grass with inflorescences in a more or less dense spike is Black-grass. However, the spike of that species is very dense and cylindrical and the leaves are blue-green. Sweet Vernal-grass is a similar perennial species that sometimes grows in field boundaries, but it has short awns.

HABITAT:
Arable and horticultural land.

SOIL TYPE:
Sandy loams and sands.

DISTRIBUTION:
Formerly present in scattered sites in Surrey and East Anglia, with a few records north to Yorkshire and Lancashire. It is now thought to be extinct.

Annual Vernal-grass: spikelet, awn ×7

LIFE CYCLE:
Flowers from June to October. Seed biology is unknown. Germination period is unknown.

J F M A M J J A S O N D

REASONS FOR DECLINE:
This species may originally have been introduced with crop seed in the 19th century, and its decline may have been largely due to better seed-cleaning.

| --- | --- | --- | --- | --- | --- | --- | --- | --- |
		POACEAE: grasses						
136	258	Dense Silky-bent	Apera interrupta	erect	10—70 cm	erect; narrow interrupted spike with stalkless branches	3—20 cm long, up to 1·5 cm wide	3—10 cm long by 4mm wide; leaf sheaths green or purple
137	260	Loose Silky-bent	Apera spica-venti	erect	20—100 cm	erect; open; spreading; much branched	10—25 cm long, up to 15 cm wide	7—20 cm long by 3—10 mm wide; leaf sheaths green or purple
138		Black Bent	Agrostis gigantea	erect, loosely tufted	40—120 cm	erect; much branched; open and loose	8—25 cm long, up to 0·8 cm wide	5—20 cm long by 2—8mm wide
139		Creeping Bent	Agrostis stolonifera	erect; tufted	8—40 cm	erect; long and narrow with parallel sides with clustered branches	1—13 cm long, up to 2·5 cm wide	1—10 cm long by 0·5—5·0mm wide
140	246	Great Brome	Anisantha diandrus	robust; softly and densely hairy	30—80 cm	nodding; much branched with erect long-stalked spikelets; very loose; green or purplish	up to 25 cm long and wide	20 cm long by 4—8mm wide
141		Barren Brome	Anisantha sterilis	robust; softly and densely hairy	30—100 cm	drooping; simply branched; green or purplish	up to 25 cm long and wide	5—25 cm long by 2—7mm wide
142	244	Field Brome	Bromus arvensis	robust; softly and densely hairy	up to 100 cm	spreading; much branched with long drooping, long-stalked spikelets; loose and open; green or purplish	8—25 cm long, up to 20 cm wide	20 cm long by 5mm wide
143	250	Rye Brome	Bromus secalinus	robust; softly and densely hairy	up to 100 cm	erect to nodding; much branched; loose with long stalked spikelets; green or purplish	5—20 cm long	20 cm long by 5mm wide
144	248	Interrupted Brome	Bromus interruptus	robust; softly and densely hairy	up to 100 cm	erect; branched; dense with short-stalked spikelets irregularly placed up the stem; greyish-green	2—9 cm long, up to 2 cm wide	20 cm long by 5mm wide
145		Meadow Brome	Bromus commutatus	erect; lower parts softly hairy, upper parts almost hairless	30—90 cm	nodding; loose and open; green or purplish	6—25 cm long	30 cm long by 3—9mm wide
146		Soft-brome	Bromus hordaceus	erect; softly hairy	5—80 cm	erect, narrow; loose to nodding; greyish-green or purplish	2—16 cm long, up to 6 cm wide	20 cm long by 2—7mm wide
147	262	Annual Vernal-grass	Anthoxanthum aristatum	erect or spreading; solitary or multiple stems	10—40 cm	spike-like; narrowly spear-shaped to oblong moderately dense; yellowish green, pale green or purplish	1—3 cm long, up to 12 mm wide	5cm long by 2—4mm wide; green; tufts of hair where leaf meets stem
148		Sweet Vernal-grass	Anthoxanthum odoratum	erect; tufted	10—100 cm	spike like; oval to narrowly oblong; dense to loose; grren or purplish	1—12 cm long, up to 15 mm wide	1—30 cm long by 1·5—9·0mm wide
149	254	Nit-grass	Gastridium ventricosum	erect; grows in tufts; hairless	10—50 cm	spike-like; oblong, spear-shaped tapering upwards to a point; pale green	2—10 cm long, 5—12 mm wide	10 cm long by 4mm wide
150		Black-grass	Alopecurus myosuroides	erect; compact or solitary; hairless	20—80 cm	spike-like; narrowly cylindrical; dense; yellowish-green, pale green or purplish	2—12 cm long, 2—8mm wide	3—16 cm long by 2—8mm wide; blue-green
151	256	Lesser Quaking-grass	Briza minor	erect; hairless; forming loose tufts or solitary	10—50 cm	erect; branched; loose	4—20 cm long, 2—10 cm wide	3—14 cm long by 3—7mm wide; blue-green
152		Quaking-grass	Briza media	erect; hairless; stiff; forming loose tufts	15—75 cm	erect; pyramidal shape; branched; loose	4—18 cm long and wide	4—15 cm long by 2—4mm wide
153	252	Darnel	Lolium temulentum	erect	30—90 cm	erect, rigid spike	10—30 cm long	6—40 cm long by 3—13mm wide
154		Italian Rye-grass	Lolium multiflorum	erect	30—100 cm	erect or nodding spike	5—20 cm long	6—25 cm long by 4—10mm wide
155		Perennial Rye-grass	Lolium perenne	erect	10—90 cm	erect, straight or slightly curved, stiff spike	4—30 cm long	3—20 cm long by 2—6mm wide

DESCRIPTION OF LIGULE	SIZE	DESCRIPTION OF SPIKELET	SIZE	DESCRIPTION OF LEMMA	SIZE	DESCRIPTION OF AWN	SIZE
long, narrow and pointed	5mm long	narrowly oblong	2·0–2·5mm long; anthers c. 0·4mm	spear-shaped	c. 2mm long	fine; straight	4–10mm long
oblong	3–10mm long	narrowly oblong	2·5–3·0mm long; anthers 1–2mm	narrowly spear-shaped	c. 2mm long	fine, straight or flexible awn	5–10mm
very blunt	1·5–6·0mm long	spear-shaped to oblong	2–3mm long; anthers 1·0–1·5mm	oval to oblong; very blunt	c. 2·5mm long	very short awn or awnless	
blunt	1–6mm long	densely clustered; narrowly oblong	2–3mm long; anthers 1·0–1·5mm	oval to oblong	c. 1mm long	usually awnless, sometimes with very short awn	up to 50mm long
blunt	up to 6mm long	rectangular, laterally compressed	30–40mm long; anthers 1–4mm	narrowly oval, tapering to a point; rough	25–30mm		awn up to 25mm long
toothed	2–4mm long	narrow, parallel-sided	20–25mm long; anthers 1mm	narrowly oval, tapering to a point; rough	14–18mm		10mm long
blunt	up to 4mm long	oval, laterally compressed; hairless	15mm long by 4mm wide; anthers up to 5mm long	bluntly angled; inrolled edges	< 6·5mm long	rough; straight	10mm long
blunt	up to 4mm long	oval, slightly laterally compressed	15mm long by 4mm wide; anthers c.1mm long	oval; inrolled edges	up to 9mm long	straight	8mm long
blunt; hairy	up to 4mm long	oval, slightly laterally compressed; hairy	10–17mm long; anthers c. 1mm	oval	7–8mm long	rough; straight	8–9mm long
hairless	1–4mm long	oval; hairless	15–20mm long; anthers 1·0–1·5mm	oval	9mm		5–8mm long
short; hairy	up to 2·5mm long	oval, tapering to a point	15–20mm long; anthers 1·0–1·5mm	oval	8–9mm long	straight	lower 4–5mm long; upper 7–10mm
pointed	up to 2mm long	spear-shaped to oblong	5·0–7·5mm long; anthers 2·5–3·5mm	lower narrowly oblong; upper elliptical	lower 4·5mm long; upper 2mm	conspicuously bent; lower lemmas with finer, shorter awns than upper lemmas	lower 2–4mm long; upper 6–19mm
blunt; long	1–5mm long	spear-shaped	6–10mm long; anthers 3–4·5mm	lower narrowly oblong; upper roundish	lower c.3mm long; upper 2mm	lower fine, straight; upper more stout, bent	3–4mm long
long	1–3mm long	densely overlapping; narrowly oblong	4mm long; anthers 0·7mm	elliptical with toothed tip	1mm long	slender; bent at the middle	9–13mm long
blunt	2–5mm	narrowly oblong	4·5–7·0mm long; anthers 3–4mm	oval	5mm long	arising from near base of lemma; straight	
blunt; long	3–6mm	nodding; triangular, shining, green tinged with purple	3–5mm long; wider than long; anthers 0·6mm	overlapping; broad at base; rounded at top		no awns	
blunt; short	0·5–1·5mm	drooping; elliptical to oval; usually purplish	4–7mm long; wider than long; anthers 2·0–2·5mm	overlapping; broad at base; rounded at top	4mm long	no awns	
blunt; membranous auricles clasping stem	2mm long	oblong; glume exceeds spikelet length	12–20mm long by 4–6mm wide	oval	6–8mm long	rough; straight	up to 20mm long
narrow, spreading auricles at base	1–2mm long	oblong; glume does not exceed spikelet length	8–25mm long; anthers 3·0–4·5mm	oblong	5–8mm	no awns	
blunt; projecting auricles at base	2mm long	oblong; glume does not exceed spikelet length	7–20mm long; anthers 3–4mm	oval-oblong, slightly pointed	5–7mm	no awns	

Mosses, liverworts and hornworts of arable fields

It would be difficult to fill a book such as this on the mosses, liverworts and hornworts of arable land. This is not because there are so few, for some 90 different kinds (out of a total of about one thousand in the British Isles) have been recorded on cultivated ground, but because we still know so little about their biology, distribution and current status.

Mosses, liverwort and hornworts, collectively known as bryophytes, are unlike flowering plants in that they have no flowers or seed but produce tiny spores or have other specialised reproductive structures. The three types are not particularly closely related but do share certain life-cycle characteristics. Arable mosses have small narrow or oval leaves with a midrib (nerve). Liverworts in arable land tend to be flat thalli, lobes resembling liver with curly margins or forming small rosettes. They lack true roots but are anchored by filamentous threads known as rhizoids. Hornworts look superficially similar to liverworts but there are some slight differences: they usually have thalli that are wavy or crisped at the edge, are greasy to look at and produce spores on horn-like appendages.

These plants are small, often ephemeral, only visible from autumn to spring and, it has to be said, are not easy to identify. When the green plants are not visible on the surface, they are likely to be hidden in the 'diaspore bank', in much the same way that the seeds of annual flowering plants overwinter in the soil. Cultivation subsequently brings them to the surface and triggers their growth.

In any one field there are likely to be no more than 20 bryophytes, usually far fewer, but even the 'typical' arable bryophyte assemblage appears to have declined in the last half-century or so. Evidence for this loss is largely anecdotal but many bryologists find they need to look further afield before finding suitable fields to explore. In particular, the hornworts are very rare, as is the attractive liverwort known as Balloonwort (*Sphaerocarpos*). Many of the typical arable mosses are increasingly confined to other disturbed places, such as tracks and quarries, as the farmed landscape become more inimical to their survival. Three mosses are included in the UK government's Biodiversity Action Plan (UKBAP) action plan: Sausage Beard-moss *Didymodon tomaculosus*, *Ephemerum stellatum* and *Weissia rostellata*.

It may come as a surprise to find that arable fields harbour not only flowering plants but also mosses which we usually associate with damp shady areas in woods. And the damp is the key to their survival, since they need moisture to help them reproduce. These primitive types of plant also have no water transport system so they usually lie on the surface acting as spongy carpets. Even though arable land is known to support a distinctive bryophyte flora, as yet we do not know much about them. Indeed, without the interest of the late Harold Whitehouse, they might have continued to be overlooked.

Finding and identifying bryophytes

Arable bryophytes occur where there is little or no competition from either the crop or from other flowering plants. The best time to see them is from late autumn to early spring on uncropped field margins, or, best of all, among cereal stubbles where they can complete their life cycle. The principal change in the countryside which has led to the decline of these plants has been the loss of stubble fields which used to be bryophyte havens. Nowadays, fields are rarely left for any period of time following harvest, but are cultivated almost immediately which effectively destroys the next generation of mosses and liverworts. Being such tiny plants, they cannot compete with the larger, more vigorous plants, including the crop, and, as a result, mosses have become confined to the edges where competition, and the effects of fertiliser and herbicides, are reduced.

It remains to be established whether or not organic farming is beneficial for bryophytes. Although chemicals are not generally used, other weed control measures may be just as harmful. Mosses and liverworts can survive in conventional farming systems, particularly in those using spring-cropping régimes and incorporating uncropped margins or headlands. Our knowledge of the value of other crop types for bryophytes is still rather limited.

The photographs on pages 268–269 show just a few of the species of mosses, liverworts and hornworts found in arable fields. Further information and help with identification can be obtained from the British Bryological Society (BBS), who have produced a very useful, more detailed guide.

British Bryological Society:
Mike Walton,
Hon General Secretary,
Ivy House
Wheelock St,
Middlewich,
Cheshire CW10 9AB

ww.bbs.org.uk

Ron Porley
English Nature Bryologist

Some liverworts and hornworts of arable fields

Common Frillwort *Fossombronia pusilla*   LIVERWORT

photograph × 6;

Glaucous Crystalwort *Riccia glauca*   LIVERWORT
5 mm across
photograph × 8;

Common Crystalwort *Riccia sorocarpa*   LIVERWORT
5 mm across
photograph × 8;

Texas Balloonwort *Sphaerocarpus texanus*
8-9 mm across   HORNWORT
photograph × 6;

Field Hornwort *Anthoceros agrestis*   HORNWORT
7 mm across
photograph × 6;

Some mosses of arable fields

Bicoloured Bryum *Bryum bicolor* MOSS
leafy stem 5 mm, sporophyte 5 mm:
photograph × 3;

Hasselquist's Hyssop *Entosthodon fasicularis* MOSS
9 mm tall
photograph × 3

Common Bladder-moss *Physcomitrium pyriforme*
9 mm tall MOSS
photograph × 3

Cuspidate Earth-moss *Tortula acaulon* MOSS
3 mm high
photograph × 3;

Common Pottia *Tortula truncata* MOSS
8 mm tall including sporophyte
photograph × 3;

Threats and opportunities

THREATS

Many of the threats currently facing the arable flora are escalations of those of the past century. These include improvements in herbicide efficacy and the development of more competitive crop varieties. In addition, there are new threats, some of which are the result of innovations in agricultural technology and others that are due to global economic changes. Owing to the speed of change, it is likely that the prospect of some of these innovations will recede, while totally unforeseen developments will arise.

One of the major problems facing the conservation of the arable flora is simply the lack of information on the distribution of even the rarest species. Without such information it is impossible to conserve populations, even with the best site protection mechanisms.

Economics, local and global

It has become increasingly apparent that agricultural practices have changed in different ways in different parts of the country. This is a result of an overall decrease in the prices of most agricultural products. In the largely arable south and east of England, fewer livestock are being kept and much grassland has been converted to intensive arable land. At the same time, arable fields in the west of the country have become less profitable, and many have been converted to pasture. These western fields are important strongholds for many nationally threatened species, including Small-flowered Catchfly, Purple Viper's-bugloss, the endemic Western Ramping-fumitory and Purple Ramping-fumitory. The zone of mixed farming that formerly occupied much of central and central-southern England has now all but disappeared, with catastrophic effects particularly on farmland birds.

Changing farming economics have also resulted in many small farmers going out of business, with their land becoming part of larger enterprises. Whilst there has not been any analysis of the relative biodiversity of different sized farms, it is likely that smaller holdings offer more opportunities for biodiversity in general and have a greater tradition of land stewardship. The agricultural recession of the 1920s and 1930s led to the widespread abandonment of land, particularly arable land. If this were to be repeated, it would have drastic effects on the arable flora.

Land abandonment can take other forms. Set-aside is a market control mechanism that encourages the withdrawal of arable land from production. Although this is usually detrimental to the arable flora, it can benefit other farmland wildlife and, if used creatively, can be an opportunity for the conservation of arable plants. The development of land for housing, industry or transport infrastructure is a more permanent change of land use, and is a particular threat at the beginning of the 21st century.

These economic changes are partly a reflection of global trends. With the increasing liberalisation of world trade, it is possible that even greater polarisation of agricultural production is in prospect.

Genetically modified crops

At the beginning of the 21st century, biotechnology is beginning to produce novel genetically modified (GM) crop varieties that can be grown and managed in different ways from existing varieties. This, potentially, has profound implications for the conservation of arable plants, whose demise is already inextricably linked to agricultural intensification. If GM crops lead to even more intensive management practices than are involved in growing conventional crops, they could prove to be the last straw for many of our endangered arable plants, as well as the organisms that depend on them for food and shelter.

GM crops which are likely to be considered for commercialisation over the next decade can be divided into two groups: those with modified agronomic traits (*e.g.* tolerance to broad-spectrum herbicides, resistance to pests and diseases, and tolerance of drought or saline conditions); and those with modified quality traits (*e.g.* altered protein, oil or carbohydrate content to cater for specific food or industrial products).

GM crops with modified quality traits are unlikely to require new management practices, except perhaps those practices relating to containment of transgenic material and segregation of harvested products from conventionally-grown products. However, GM varieties with high added value to the farmer could potentially replace other, less-profitable crops in a rotation, and/or be grown in areas where that crop was previously uneconomic. This could have management implications for the whole rotation, although such changes can apply to all new crop introductions and not just to GMOs. Genetic modification could lead to a change towards more environmentally-damaging cropping systems. For example, if a GM Maize could be grown in more northerly parts of the UK, it could replace traditional forage grass areas that are important for plants, insects and birds.

GM crops with altered agronomic traits will, by definition, lead to changes in management. However, the impacts on biodiversity may be hard to predict without actually growing the novel crops on a field-scale and scientifically monitoring the effects on wildlife.

GM herbicide-tolerant (HT) crops are among the first to have come up for commercial consent in Europe. The principal varieties which could potentially become available to UK farmers are Maize, winter and spring Oilseed Rape tolerant to glufosinate-ammonium ('Liberty'), and Sugar and Fodder Beet tolerant to glyphosate ('Roundup'). These crops are designed to withstand over-the-top applications of broad-spectrum herbicide during the growing season, which represents a significant change in management from conventional varieties. Such systems could damage arable plant biodiversity, and therefore also the diversity of invertebrates, birds and mammals associated with arable plants, for the following reasons:

▸ Broad-spectrum herbicides are highly efficient at killing a wide range of plant species, so plant diversity in and around cropped areas is likely to be reduced.

- Over-reliance upon a single herbicide could eventually lead to weed shifts towards more resistant varieties, ultimately worsening current problems of resistant weeds.

- Spraying fields during the growing season can increase the likelihood of spray drift damage, since spray nozzles need to be set high, and marginal and hedgerow plants will be in full leaf. This could cause significant damage to wild plants in these habitats.

- The creation of a cropping system designed to rely on herbicide inputs is a significant move away from current efforts by conservation groups to promote more careful and restricted use of herbicides. Seeds of GM crops will cost more than conventional varieties, so farmers will naturally want to exploit their novel properties to the full.

However, it is also possible that GM herbicide-tolerant crops could offer greater flexibility in weed control and, if planted judiciously, could benefit arable plant biodiversity. For example, spraying could be delayed until later in the growing season, allowing early-germinating arable plants to survive and set seed. It is also possible that the need for 'insurance' applications of pre-emergence herbicides would be reduced. In addition, on lighter soils no-till farming might also be possible, with chemical control being used instead of ploughing. However, it is unclear whether such practices would offer net benefits to arable plant diversity. More studies are needed to determine whether herbicide-tolerant crops will be harmful or beneficial to farmland wildlife. Whatever the outcome, it is likely that, if any GMHT crops are commercialised, they will be subject to restrictions in management practices to prevent harmful impacts on our countryside.

Anna Hope
ENGLISH NATURE BIOTECHNOLOGY UNIT

OPPORTUNITIES

During the 1990s there was a growing willingness among the farming community to consider measures for the conservation of farmland biodiversity, including arable plants. Most farmers have always been very interested in the wildlife on their land, but until relatively recently they have been encouraged to maximise production at any cost. However, with the aid of agri-environment schemes, farmers can now be compensated for the income forgone when undertaking ambitious conservation programmes.

Action plans

At the same time, conservation organisations have recognised the importance of the wildlife found on arable land. The inclusion of arable species in the UK Biodiversity Action Plan (Appendix 3), and the development of a Habitat Action Plan for cereal field margins, have been extremely important in this respect. They have raised the profile of these formerly neglected species and are helping to ensure that their needs are catered for within agri-environment schemes. Many Local Biodiversity Action Plans (LBAPs) now include a section on the conservation of arable wildlife.

The voluntary approach

Much conservation work is carried out on farms without any form of financial aid. The Farming and Wildlife Advisory Group (FWAG) provides advice on the management of farmland for wildlife, and has had much influence especially on the management of field boundaries and the less productive areas of farms. The Game Conservancy Trust pioneered research into the ecology and conservation management of arable land and field boundaries during the 1980s. One of the early fruits of this research was recognition of the value of 'Conservation Headlands' (see page 284). Originally designed to improve the food supply for gamebird chicks, Conservation Headlands were seen to have benefits for other farmland wildlife, including arable plants. They were widely adopted by farmers, and have become an integral part of some Countryside Stewardship options and several Environmentally Sensitive Areas (ESAs).

Agri-environment schemes

In England, Environmentally Sensitive Areas and the Countryside Stewardship Scheme have been providing financial support for the conservation management of farmland since 1987 and 1991, respectively. These are administered by the Rural Development Service of The Department for Environment, Food and Rural Affairs (Defra). However, until recently, the impact of these schemes on arable land has been limited, although under certain circumstances cultivated but uncropped field margin options were available for the management of arable plants. In 1998, an Arable Stewardship Scheme was introduced in pilot areas in East Anglia and the West Midlands. This was aimed specifically at the management of arable land and offered much for the conservation of arable plants, in particular, through the uncropped margin option. Measures from this pilot scheme have now been introduced throughout England. The Entry Level Scheme in England was introduced in April 2003. It aims to enable farmers across a large part of the country to deliver simple but effective management on their farm. The Scheme seeks to address issues such as diffuse pollution, loss of biodiversity, landscape character and damage to the historic environment. The scheme is being piloted in Barnard Castle, Co Durham; Market Deeping, Lincs; Tiverton, Devon and Mortimer in Berkshire. See website for more details:

http://www.defra.gov.uk/erdp/reviews/agrienv/qanda.htm

Wales

In Wales, Tir Gofal is managed by the Countryside Council for Wales on behalf of the National Assembly of Wales. The scheme is funded from the Welsh Assembly budget and the European Union through EC Regulation 1257/99. It is a whole-farm scheme and is available to farmers throughout Wales.

There are specific provisions for the management of arable land which include payments for leaving cereal crops unsprayed, retaining winter stubbles and undersowing spring cereals.

Northern Ireland

Northern Ireland has three agri-environment schemes: The Environmentally Sensitive Areas' Scheme, The Countryside Management Scheme and the Organic Farming Scheme.

Northern Ireland Agri-Environment Schemes' Arable Options

Arable farmland constitutes less than 4% of farmed land in Northern Ireland. Arable options are available as part of the Countryside Management Scheme (CMS), which applies to land outside designated Environmentally Sensitive Areas in Northern Ireland. Since 2000, entrants in the Environmentally Sensitive Area (ESA) Scheme have also been able to choose to participate in the following arable options.

CMS and ESA Arable Options – 'Arable fields managed for wildlife':
- Retention of winter stubbles.
- Conversion of improved grassland to spring cereal or oilseed rape.
- Planting wild bird cover on improved grassland.
- Planting wild bird cover as an arable crop margin.
- Creation of a rough grass field margin.
- Establishment of a conservation crop margin.

Details of the Countryside Management Scheme and the Environmentally Sensitive Areas Scheme can be obtained from the Department of Agriculture and Development for Northern Ireland (DARDNI), Countryside Management Division, Dundonald House, Belfast (see page 305).

Scotland

The Rural Stewardship Scheme (RSS), which is part of the Scottish Rural Development Plan, provides assistance to encourage farmers, crofters and common grazings committees to adopt environmentally friendly practices and to maintain and enhance particular habitats and landscape features. The scheme is discretionary, with all applications being subject to a selection process through a ranking system. Applicants undertake to participate in the Scheme for a period of at least 5 years. In addition, they must agree to manage the relevant areas of land and carry out appropriate capital works in line with the rules and conditions of the Scheme. Certain general environmental requirements (Standard of Good Farming Practice and General Environmental Conditions) also apply to the farm, croft or common grazing as a whole and not just to those areas or features that are being positively managed under the RSS.

The RSS currently includes a number of management options which are particularly relevant to arable farms:

- Management of Grass Margins or Beetle-banks in Arable Fields.
- Management of Conservation Headlands.
- Management of Extended Hedges.
- Management of Hedgerows.
- Introduction or Retention of Extensive Cropping (Less Favoured Areas only);
- Management of Cropped Machair.
- Unharvested Crops.
- Spring Cropping (non-Less Favoured Areas only).

The Scottish Executive intend to bring the area of land in Scotland covered by agreements under agri-environment schemes to 2,000,000 hectares during 2003–2004. For more information on the current scheme, please refer to the RSS Explanatory Booklet. You can access this in *pdf* format on the Scottish Executive web-site at:

http://www.scotland.gov.uk/library3/environment/rss-00.asp

Agri-environment schemes may come and go, and there is no guarantee that they will be available at all in the future. For the long-term survival of a species-rich arable flora in Britain, it will be essential to integrate its conservation into lower-intensity arable systems, and to prove the intrinsic worth of arable plants within healthy farming ecosystems.

Minimising inputs – Integrated Crop Management
The application of pesticides and fertilisers is becoming much more efficient and there is less wastage of agrochemicals in non-target areas. There is also a trend away from prophylactic spraying, towards Integrated Crop Management (ICM) systems, in which labour and materials are saved by avoiding unnecessary weed control and fertiliser application. ICM systems also make use of buffer strips to minimise chemical and fertiliser run-off, and nectar crops and beetle-banks to encourage beneficial insects, thereby cutting insecticide bills. For the conventional farmer, ICM makes economic and environmental sense.

Organic farms
Taking long-term agricultural sustainability a step further, organic farming is becoming much more popular with both producers and consumers. Organic methods use no herbicides or other pesticides, and neither do they use artificial fertilisers. They do, however, use green manures, crop rotations and mechanical weeding to eliminate non-crop plants, and indeed there is even more incentive to control non-crop plants than in a conventional system where there is ready recourse to herbicides if necessary. Studies both in Britain and in other European countries have shown that arable plant diversity is significantly higher on organic farms than on conventionally managed farms.

The Management of Arable Land for Plant Conservation - Practical Guidelines

Introduction

To be effective, management actions to conserve arable plants should be inexpensive and easy to carry out, but, above all, must fit into normal farming practice. In many cases, management will be straightforward, although some arable plants do have individual needs. Also, if the land is not farmed in a conventional manner, then management must be specifically designed. In these cases, advice can be obtained from English Nature, the Countryside Council for Wales, Environment and Heritage Service Northern Ireland, Scottish Natural Heritage, the Rural Development Service (RDS) of Defra, RSPB, Plantlife, FWAG, or The Game Conservancy Trust, or from a reputable ecological consultant.

Finding the right fields

Land converted to arable within the last 100 years, or which is subject to repeated and heavy herbicide and fertiliser use stands little chance of supporting uncommon arable plants. Some areas of the country are richer for arable plants than others, and there is a greater chance of success in restoring arable communities in these areas (see map on page 17). Heavy land is also less likely to support arable plants than fields on light soils. To find out whether an area of land supports uncommon arable plants, information can be sought from local botanists, country agencies, Defra, local Wildlife Trust, County Environmental Records Centre or local BAP officer.

Features for selecting areas for management for uncommon arable plants	
Continuity of management	Sites with a long history of arable management (in cultivation before the 1840s) are thought to be richer in uncommon species than fields only recently cultivated. Consult tithe maps (compiled in the 1840s and 1850s) at your local county records office.
Soil	In general, lighter soils have a richer flora than heavy or very fertile soils (there are of course exceptions). Thin soils at the top of a slope are usually better than the deep, accumulated soils at the bottom.
Aspect and shading	Heavily shaded field margins are usually less good than exposed, sunny, south-facing margins.
Problem species	Field margins with serious infestations of Cleavers, Barren Brome and Black-grass are usually poor in uncommon species. Do not choose field margins along recently established tracks.

Tailoring management for specific plants

Each arable plant species has specific needs (see individual species accounts for details). Knowing how each plant behaves and responds to management operations can make the difference between a thriving arable plant population and an extinct one. The following are some useful pointers:

- Most arable plants are susceptible to broad-spectrum herbicides.
- Most uncommon arable plants compete poorly with fully fertilised crops.
- A few, like Cornflower and Field Gromwell are capable of growing within a modern crop canopy.
- A few plants, like Grass-poly require seasonally waterlogged conditions.
- Arable plants do better within a crop drilled at the same time as they germinate (see table below for some examples).
- A late-flowering arable plant may fare better where stubbles are left after harvest.

Autumn/winter-germinating arable plants	Spring-germinating arable plants	Plants that can germinate in both spring and autumn
Corn Buttercup	Corn Marigold	Rough Poppy
Shepherd's-needle	Weasel's-snout	Broad-leaved Spurge
Mousetail	Red Hemp-nettle	Pheasant's-eye
Cornflower	Small-flowered Catchfly	Narrow-fruited Cornsalad

Some late-flowering arable plants:
Corn Parsley, Red Hemp-nettle, Night-flowering Catchfly, Round-leaved Fluellen, Sharp-leaved Fluellen

Site characteristics
Most arable plant sites will be arable field margins on conventionally-managed farms. A few farms will have an organic system, which will pose its own constraints. In a very few cases, the site will be part of a reserve or a Site of Special Scientific Interest (SSSI). Some sites will be in semi-permanent grassland, on road verges or in set-aside. Small, or awkwardly-shaped, fields may have parts that wide-boom sprayers cannot reach. These different conditions will all influence the range of management options that are available.

Organic systems
Organic systems do not use herbicides, but control weeds by crop rotation and cultivation. Weed control in these systems can sometimes be very efficient. Arable plants are not confined to field edges in these systems, but are frequently distributed throughout the field. Although organic systems will be less inimical to uncommon arable plants, special measures, including uncropped wildlife strips, may still be required.

Management in non-arable sites
Uncommon arable plants can sometimes occur in non-arable situations, including eroding clay cliffs, coastal shingle, parched grasslands, grass leys,

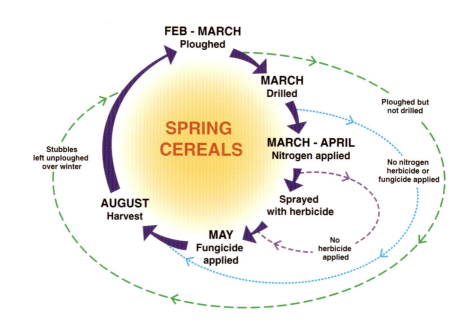

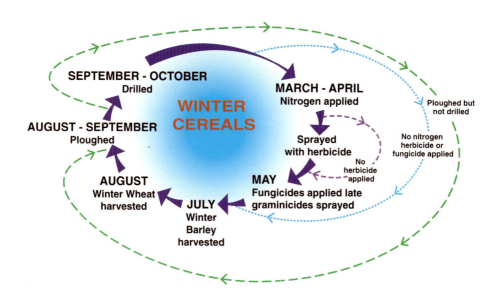

Annual management cycles for spring and winter cereal crops

Some examples of arable species growing in non-arable habitats

Small-flowered Catchfly	Eroding clay cliffs near Bournemouth.
Corn Parsley	Parched grassland on sea walls and cliff-tops.
Red Hemp-nettle	Coastal shingle in the south-east and east of England Tracks on former arable land on Salisbury Plain.
Red-tipped Cudweed	Parched acidic grasslands in Suffolk and Surrey.
Broad-leaved Cudweed	Chalk quarries in Surrey, Sussex and Oxfordshire.
Pheasant's-eye	Tracks on former arable land on Salisbury Plain and in Sussex.

tracks and road verges. As in arable field margins, the aim is to maintain open conditions free from competition with other plants. Naturally open habitats, including cliffs and shingle, are essentially self-maintaining. The grazing régime of parched grasslands is vital for maintaining open conditions, and Rabbits frequently play a key rôle in these régimes. Where sites are liable to be overgrown by more competitive vegetation, such as in a newly established grass ley or on a road verge, periodic cultivation should be carried out.

Management options

Which part of the field to manage?
In most fields, any remaining species-rich seedbank will be restricted to field edges, so arable plant management will normally be focussed here. The rest of the crop can be managed normally. A six-metre margin strip is often both the most convenient area for the farmer and the best area for plants. Organic, or other low-intensity régimes, such as summer fallow areas managed for Stone-curlew, can enable rare arable plants to occur throughout the field.

Field margins are usually managed in one of three ways:

1. *Conservation Headlands* are field margin strips sown with a crop along with the rest of the field and fully fertilised. Only highly specific herbicides are used, and there are restrictions on fungicide and herbicide use.
2. *Low-nitrogen Conservation Headlands* are similar to Conservation Headlands but without nitrogen applications. Many farmers would not consider it worthwhile to apply any other inputs to these field margins. Field margins of this type are similar to those extensively used in Germany, and have proven very successful in the Arable Stewardship Pilot Areas.
3. *Uncropped Cultivated Margins* have been supported in the Breckland Environmentally Sensitive Area (ESA) for several years, were included in the Arable Stewardship pilot and are now available more widely in the Countryside Stewardship Scheme. For most arable plants, Uncropped Cultivated Strips are the optimum form of management. The field margin strip is normally cultivated at the same time as the rest of the field, but no crop is drilled and no agrochemicals applied.

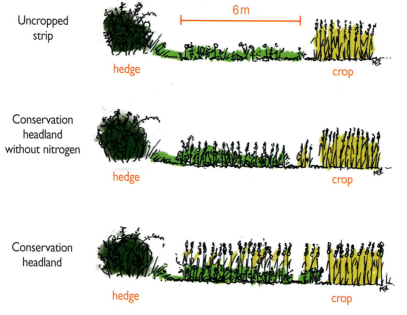

	6 m
Uncropped strip	hedge ——— crop
Conservation headland without nitrogen	hedge ——— crop
Conservation headland	hedge ——— crop

Cross-sections of field margins under different forms of conservation management

Cultivation times

In most cases, field margin strips are cultivated with the rest of the field. However, this is not ideal for those arable plants with short-lived seed, and cultivation times should be altered to suit the arable plants present. For example, if Corn Buttercup is present in a field that is to be drilled with spring Linseed, then the margin strip can be cultivated in the preceding autumn and left uncultivated in the spring when the rest of the field is ploughed. If a spring-germinating plant such as Red Hemp-nettle is present in a field which is to be drilled with winter Wheat, then cultivation in the spring would be beneficial. Optimum cultivation times for each arable plant are given in the individual species accounts.

What crop to grow?

The aim is always to reduce competition from the crop to a point at which the arable plants can flourish. Different crops exert different amounts of competition. A dense cereal crop, for instance, will be more competitive than a crop of Sugar Beet. Where the crop is not fertilised, differences between crop types become less important. On conservation headlands with nitrogen, winter Barley is more competitive than winter Wheat, and spring Barley is more competitive than spring Wheat. Oats and Maize are very competitive crops, whilst Linseed, root and vegetable crops are much less so. Drilling time is also an important factor which can determine the arable plants that germinate. Root crops and Maize are

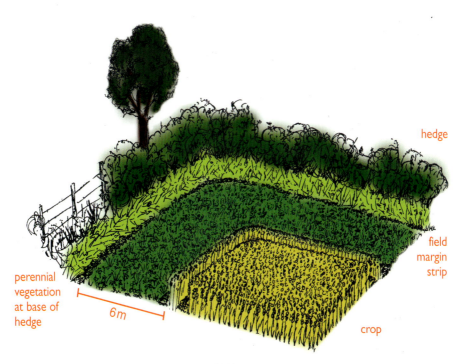

Field corner

typically drilled late in the spring after most of the less common arable plants have germinated, and often support a very specialised flora.

Problem species

The greatest weed problems in modern arable farming are Black-grass, Barren Brome and Cleavers. They respond well to nitrogen, can compete with modern crop varieties and germinate in the early autumn, producing seed well before harvest time. They are also difficult to eradicate with herbicides. Fortunately, a switch to lower input régimes in field margins will cause these nitrogen-demanding weeds to decline over time.

A number of other perennial weeds can take advantage where there is no weed control and little competition. These can seriously affect numbers of uncommon arable plants. Most of these perennials have rhizomes that grow beneath ploughing depth (*e.g.* Creeping Thistle and Perennial Sow-thistle), or have tubers that are broken up by cultivation into small pieces each of which can form a new plant (*e.g.* Couch and Onion Couch).

These weeds can be controlled either by cultural methods or herbicides. Glyphosate has been successfully used for control of Perennial Sow-thistle by the National Trust in Cornwall, and clodinafop-propargyl has been effective in controlling Onion Couch at the Somerset Wildlife Trust reserve at Fivehead. The weeds listed can all be controlled, and eventually eliminated, by continuous

Herbicides with potential for the control of problem species

Species	Persistent part	Herbicide
Couch	Rhizomes that break up on cultivation.	Glyphosate in autumn after harvest and seed production by most rare species but while the weather still allows plant growth.
Onion Couch	Bulbs that break up on cultivation.	Spray clodinafop-propargyl or flamprop-isopropyl in late spring/early summer.
Creeping Thistle	Rhizomes below plough depth.	Spot-treatment in mid-summer with glyphosate or clopyralid using a weed-wipe.
Perennial Sow-thistle	Rhizomes below plough depth.	Spot-treatment in mid-summer with glyphosate or clopyralid using a weed-wipe.
Colt's-foot	Rhizomes below plough depth.	Spot -treatment with glyphosate or clopyralid using a weed-wipe.
Bramble	Rhizomes below plough depth and buds that are dispersed by cultivation.	Spot-treatment with glyphosate using a weed-wipe.

grazing or cutting, although this must sometimes be carried out for several years. This approach can itself cause problems for uncommon arable plants, and may be the last straw for species with poorly persistent seedbanks. Much further work is required on the effects of competitive perennial weeds on arable plant communities and on methods for their control.

Intensive farmland adjacent to Old Winchester Hill, Hampshire.

Managing arable flowers

Agri-environment schemes offer many opportunities for land-owners to conserve arable plants, and there are now several examples of protected sites and reserves. Twenty nine species of arable plants are included in the UK Biodiversity Action Plan (BAP) (Appendix 3). This chapter describes some current conservation initiatives in both farmed and non-farmed situations.

Protected sites

Nine arable sites in England have been given protection as Sites of Special Scientific Interest (SSSI) under the Wildlife and Countryside Act, or are managed as a nature reserve specifically for their arable plant interest. A further 50 SSSIs support some arable plants. The guidelines on notification of SSSIs do include arable land (NCC, 1989) but the majority of BAP plant populations and other scarce and declining arable plants are situated on the field margins of undesignated farmland.

In 1985, the first arable SSSI was scheduled near Rochester in Kent. Other arable SSSIs that have been designated include a large area in Cambridgeshire where the very rare Grass-poly grows in areas that are waterlogged in the winter, several sites in the Breckland of Norfolk and Suffolk where rare species typical of that area grow, and a group of three fields in mid-Somerset.

Arable Sites of Special Scientific Interest		
County	**Location name**	**Species found on site**
Cambridgeshire	Whittlesford Hummocky Fields	Grass-poly
Hampshire	Lake Allotments, Isle of Wight	Martin's Ramping-fumitory Purple Ramping-fumitory
Kent	Cobham Woods	Ground-pine Rough Marsh-mallow
Norfolk	Weeting Heath	Breckland Speedwell Fingered Speedwell Spring Speedwell
Somerset	Fivehead Arable Fields	Broad-fruited Cornsalad Corn Buttercup, Broad-leaved Spurge Spreading Hedge-parsley
Suffolk	Cherry Hill and the Gallops	Breckland Speedwell
	Potton Hall Fields	Red-tipped Cudweed
Surrey	Hankley Farm	Red-tipped Cudweed
	Fames Rough	Broad-leaved Cudweed Ground-pine

Conservation and game management – conservation headlands

The aim of conservation headland management is to provide a resource of broad-leaved plants within the arable crop on which the insects important for newly-hatched gamebird chicks can feed. Extensive research has been undertaken to ensure that such management does not compromise crop yields or quality and does not lead to the increase of troublesome weeds such as Black-grass and Cleavers. Conservation headland management is carried out on the outermost 6 m of the cropped area of the field. The crop is drilled and fertilisers and fungicides are applied as usual. The only insecticides permitted are highly selective aphicides for the control of Barley Yellow Dwarf Virus (BYDV).

Broad-leaved weed herbicides are not applied, but selective compounds for the control of Wild Oats and Black-grass are permitted. Other herbicide treatments are recommended in particular circumstances.

The omission of non-specific herbicides can permit the survival of a wide range of arable plants, although the application of full levels of fertiliser may mean that only the more competitive and shade-tolerant species survive.

Guidelines for the management of conservation headlands and associated field boundaries can be obtained from the Game Conservancy Trust. A field officer can provide free on-farm advice.

Ackerrandstreifenprogrammen – arable plant conservation in Germany

The first field margin strips managed for the conservation of the arable flora were in the Eifel Mountains of Germany in the late 1970s. These resulted from the dedication and hard work of Dr Wolfgang Schumacher of the University of Bonn. He realised that Germany's traditional arable flora was rapidly disappearing under the tide of agricultural intensification during the 1970s. Species including Summer Pheasant's-eye, Lesser Bur-parsley and Lamb's Succory were rapidly approaching extinction over much of the country. He realised that there was little hope of attracting government funding for the conservation of 'weeds' until it had been shown what was possible.

He therefore persuaded farmers in the Eifel Mountains to set up a pilot scheme.

Nitrogen and herbicide applications were omitted from between 15 km and 20 km of field margins between 1978 and 1981. This was so successful that field margins were managed with government funding in all FDR Länder by 1986, and in 1991 all Länder in the whole of the unified Germany were included. These field margin strips are not only successful in conserving arable plants, but have also been very popular with farmers and the public.

Other arable conservation initiatives that have been introduced in Germany include open-air agricultural museums and areas where traditional arable plant communities are maintained in cultivation. Some areas with rich floras are managed as reserves.

Some county Wildlife Trusts have now listed important arable land as County Wildlife Sites or Sites of Nature Conservation Importance. These designations have no force in law, and there is no funding to support changes in management, but they can be important in targeting grant-aid from agri-environment schemes and in encouraging voluntary conservation efforts by landowners. The National Trust became concerned about the conservation of endangered arable plants on their land in the late 1980s. Favourable management was introduced to several holdings on the coast of Cornwall in the mid-1990s and one site, near Newquay, is now managed as an arable nature reserve.

The wider countryside

In Europe, changes in the arable flora have mirrored those in Britain. Many species have declined throughout their ranges, and some are critically endangered. In Germany, however, the observed declines have been acted upon to great effect and can serve as a model for other countries to follow. Since 1992, funding has been available for conservation schemes in arable habitats throughout the European Union.

SUMMARY

To look after arable flowers

▶ Find out which arable plants grow on your land - contact local botanists for help and more information.

▶ Establish arable flower areas.

▶ Cultivate and harvest arable flower margins at the same time as the rest of the field.

▶ Leave stubble for as long as possible before ploughing and sowing the next crop.

▶ Seek financial assistance for arable plant margins (see pages 273–275).

Remember to:

▶ Not use herbicide in arable flower conservation areas.

▶ Try and reduce herbicide use elsewhere and only apply after seeking advice to help protect and arable plants which might be present.

▶ Avoid the use of fertilisers which damage arable flower conservation areas and, if possible, use lower rates of fertiliser elsewhere.

▶ Avoid sowing a crop where arable flower conservation is being carried out, or sow just a thin crop.

▶ KEEP UP THE GOOD WORK IF YOU ALREADY HAVE A VARIETY OF ARABLE FLOWERS!

Conservation of farmland birds

Agri-environment schemes and set-aside can also be used for the conservation of farmland birds, and specific options have been tested in the Arable Stewardship Pilot and are now available more widely in England. Many of these options can be integrated with conservation management for arable plants. The following measures are of benefit to birds on arable farmland:

- Low input, spring-sown cereals.
- Over-wintered crop stubbles.
- Bare fallow nesting plots (see below) in arable crops.
- More targeted use of pesticides and fertilisers.
- Improved management of set-aside.
- Improved management of field margins.
- Re-introduction of limited amounts of arable land into pasture-dominated areas.
- Re-introduction of limited amounts of grassland into arable-dominated areas.
- Re-establishment of hedges and other field boundaries.

All of these measures will be of general benefit to arable plants, although spring cereals should not replace winter cereals where communities of autumn-germinating arable plants are known to occur. Grassland should not be established in field margins where there is known arable plant interest.

Special projects aimed at the conservation of Stone-curlew and Lapwing in East Anglia and Wessex and Cirl Bunting in south Devon have been set up by the RSPB in association with the Rural Development Service (RDS).

Stone-curlew/Lapwing plots
Within the breeding range of the Stone-curlew in central-southern England and East Anglia, spring-cultivated plots unsown with crops in arable fields are promoted by the RSPB to encourage this species and Lapwing to nest. These plots can be established on set-aside or grant-aided under the Countryside Stewardship Scheme. Uncommon arable plants including Red hemp-nettle and Shepherd's-needle have been recorded in several of these plots.

Field margins for Cirl Bunting
The Cirl Bunting is now confined to the coastal strip of south Devon. Numbers had fallen to a very low level by the early 1990s.
A combination of cultivated, weedy field margins and over-wintered stubbles under the Countryside Stewardship Scheme has resulted in a significant increase in the population of this bird. These field margins have also benefitted arable plants and Small-flowered Catchfly, Broad-fruited Cornsalad, Few-flowered Fumitory, Small-flowered Buttercup and Slender Bird's-foot-trefoil have all been recorded.

Stone-curlew (left) and Cirl Bunting (right)

▶ *Grey Partridge*

These fields form part of a small tenanted farm to the south west of Cambridge. They were scheduled as an SSSI by English Nature for the large population of Grass-poly which was first discovered here in 1957. Broad-leaved Cudweed was recorded until the 1970s, and other uncommon species including Night-flowering Catchfly are present. In addition to the arable plants, the site is rich in birds including Reed Bunting, Linnet, Grey Partridge and Skylark. The very rare Fairy Shrimp occurs in the seasonally-flooded areas.

The soils are mainly stony clay loams overlying chalk. The area was probably tundra during the last Ice-age, and this resulted in the formation of features known as 'pingoes', shallow hollows created by frost activity. Unusually, these have persisted as winter-flooded ponds, carpeted with a thick growth of spring-germinating annual plants including Grass-poly in the summer.

The land is managed conventionally for arable crops, with the exception of approximately 3 ha which is managed for wildlife. The SSSI management agreement requires the farmer not to use herbicide in the areas with rare arable plants, but he also entered the Arable Stewardship Pilot Scheme, and has areas managed as beetle-banks, over-wintered stubbles, unfertilised field margins and sown with wildlife seed mixtures. Some problems have been encountered with Black-grass and Creeping Thistle in unfertilised field margins, but a selective Black-grass herbicide is used where necessary.

The farmer has a good relationship with the local English Nature staff, and receives advice on management from English Nature and the Rural Development Service. He is keen to continue managing for wildlife, including arable plants.

Grass-poly

initiatives

Whilst it is vitally important to note where our native rare arable plants are growing, and to endeavour to safeguard their future, there have been a number of initiatives using seed of known local provenance to help bolster both the supplies of seeds available and the chance for people to see these flowers. In the UK, we have become used to bright green, agriculturally improved, fields but it is the duller green unimproved fields which provide the greatest plant diversity.

The Somerset Wildlife Trust reserve at Fivehead consists of three small arable fields totalling 25 ha on the south-facing slope of a low ridge of limestone and calcareous clay to the south east of Taunton. It has long been known as a good site for rare arable plants including Broad-fruited Cornsalad, Spreading Hedge-parsley, Corn Buttercup and Broad-leaved Spurge. The previous farmer had managed the fields with few herbicide inputs, largely to produce animal feed. The fields are difficult to manage, being on very heavy clay soils with many large stones and emerging springs. There is a very narrow window of time in the early autumn when crops can be sown before the fields become waterlogged. Similarly, in the spring, the wet soils dry rapidly and bake hard. Most of the neighbouring arable land has been converted to pasture,, and the site was purchased by the Trust in 1992, the year after the farmer retired.

In the year between the farmer retiring and the Trust taking over management, the fields had not been ploughed. This allowed Onion Couch and Black-grass to increase dramatically throughout the reserve. Following experimental trials, the fields were sprayed in 1996 and 1997 with the highly selective herbicide clodinafop-propargyl. This controlled the grasses sufficiently to enable successful crop establishment, and a rotation consisting of two years of winter cereals, one year of spring cereals and one year of fallow was initiated. Despite the difficulties encountered at this site, experimental management has proven successful, and it is now an excellent example of the conservation of arable plants.

Corn Buttercup

Broad-leaved Spurge

i   *Flora locale* **Flora locale was established to promote and press for the use of local provenance seed in habitat restoration projects in the UK and elsewhere in Europe. This is as important for arable plants as any other semi-natural habitat. An information and education officer will join this small organisation in 2003, working with growers and users of seed to further Flora locale's aims. More information can be obtained from the Flora locale Officer (see page 306).**

West Pentire Farm at Polly Joke in Cornwall

West Pentire Farm was purchased by the National Trust in 1960. There are 11 arable fields totalling 16 ha on an exposed headland surrounded by herb-rich maritime grassland. On the death of the tenant in 1989, it was decided to manage the land 'in hand' to ensure the protection of the arable flora. In 1994, it became the first arable area to be awarded a grant under the Countryside Stewardship Scheme. West Pentire has populations of a number of locally and nationally rare arable plants, including Venus's-looking-glass, Rough Poppy, Shepherd's-needle, Small-flowered Catchfly and Western Ramping-fumitory. There are dramatic displays of Common Poppy and Corn Marigold, and large numbers of visitors visit the site purely to view the flowers against the backdrop of the sea.

The soils are very thin and stony, and most fields are ploughed only shallowly. A power harrow has also been tried. One problem in the early years was the large amount of dead vegetation that clogged up the plough. This was overcome by a light topping and occasional winter grazing. In the first few years, no crop was planted but now a thin crop of Barley is sown to act as a winter seed source for the many birds that use the site: these include Corn, Snow and Lapland Buntings and Grey Partridge.

The biggest management problem is caused by aggressive weeds such as Couch, Perennial Sow-thistle, Spear Thistle, Creeping Thistle and Ragwort. These are encouraged by the shallow ploughing. Glyphosate is applied approximately once every 5 years on a rotational basis. It is applied in the autumn after the non-target annual species have shed their seed, and appears to have been very successful in controlling the target weeds. The populations of the desirable species seem to have responded positively.

Poppies and Corn Marigolds at West Pentire

ARABLE PLANTS AS AN EDUCATIONAL RESOURCE:

The Cornfield flower project at Silpho in North Yorkshire

Several uncommon arable plants still survive within the North York Moors National Park, at the northernmost limits of their ranges. With assistance from the National Park Authority, a 10 ha field near Silpho, Hackness, was bought in April 1999 by the Carstairs Countryside Trust, which runs the project in partnership with the Authority, local naturalists and the Ryedale Folk Museum. The aim of the project is to provide a refuge for local arable flowers and to demonstrate to others what can be achieved. The field has been divided into two: the northern half has been farmed with minimal inputs and the southern half farmed commercially, although with conservation headlands. This pattern of management aims to ensure an income flow to the Trust.

Seed has been collected from a number of uncommon arable plants, both within the National Park and beyond (including species which are now extinct locally). In 2001, the Ryedale Folk Museum at Hutton-le-Hole offered the use of an arable field at the Museum in which these species could be grown to demonstrate

what the fields might have looked like before the advent of modern farming methods. This field provides the project with a 'halfway house' before any reintroduction of plants into the field at Silpho. This has created much interest amongst visitors to the museum.

A sufficiently large seedbank of a few species is now available for broadcasting at Silpho. Species will be selected on the basis of local provenance and soil suitability, and their survival and spread within the field will be monitored.

Ryedale Folk Museum arable field

Landlife **Landlife has been using annual wildflower species on a large scale and taken rare wildflowers into the heart of urban areas. In 2002, Landlife received a Biodiversity Grant from English Nature to sow Cornflower and Corncockle on farmland to:**

‣ **manage them for seed production for use in other conservation projects;**
‣ **provide habitat for bird species; and**
‣ **give people a chance to see these rare wild flowers.**

Already, bird numbers have increased and the plants have provided a source of inspiration for a wide range of people keen to start their own wildflower projects. More information can be obtained from Landlife (see page 306).

Case study 5

Cridmore Farm on the Isle of Wight

Cornflower

Cridmore Farm is situated near Godshill in the south of the Isle of Wight. The southern part of the farm is on relatively acidic, sandy loam soils. This is probably the best site for arable plants on the Isle of Wight, and rare species that have been recorded include Small-flowered Catchfly, Cornflower, Corn Parsley, Corn Marigold, Narrow-fruited Cornsalad and, most notably, Broad-fruited Cornsalad. In some years, the Cornflower and Corn Marigold put on a most impressive display in mid-summer. The farm is also very important for other wildlife, with an extremely valuable SSSI wetland and woodland. The arable land has breeding Corn Bunting and Yellowhammer.

The arable land is managed conventionally, apart from the network of field margins which are part of a Countryside Stewardship agreement. The field margins are managed both as uncropped cultivated and unfertilised margins. Some of the field margins are managed by the farmer under a separate voluntary arrangement. With the help of advice from the Game Conservancy Trust, RDS and RSPB, the farmer has reported few agronomic problems as a result of the conservation management. He is happy to continue to manage the site in a way that is sympathetic to conservation.

Case study 6

Worth and Compact Farm in Dorset

Rough Poppy

The Isle of Purbeck in south-east Dorset has long been known as an area rich in unusual arable plants. Soils near the coast are thin, brashy clays, freely-draining and very rich in lime. The climate is warm and sunny in the summer and mild in the winter, but rainfall is high and winds can be strong, with salt spray from the sea.

Worth and Compact Farm includes several arable fields separated from the sea by only a few hundred metres of permanent grassland. These fields have populations of Pheasant's-eye, Rough Poppy and Field Gromwell. The farm also includes a Bronze Age field system with limestone grassland, some of which is scheduled as an SSSI.

The most important field is managed under the Countryside Stewardship Scheme, with additional support from Purbeck District Council through the Purbeck Biodiversity Project. This field is surrounded by a 6m uncropped margin which is cultivated in the autumn and then swiped at the end of the following summer. Adjacent to this is a 12m wide conservation headland which is drilled with crop seed and fertilised. The only pesticide applied to this conservation headland is a specific Wild Oat herbicide.

The field has only been managed for its arable plants since autumn 2000, and it is therefore too early to say whether management has been successful. The farmer has encountered no problems in managing this land for conservation, and this should continue into the future.

AN ORGANIC FARM:
Bushey Ley Farm in Suffolk

Corn Buttercup

Bushy Ley Farm is situated to the north west of Ipswich in Suffolk, on the gently undulating boulder clay plain. The soils in this area are very heavy, and the climate is relatively continental, with warm, dry summers and cold winters. The area is ideally suited to cereal cultivation, and the surrounding land is intensively farmed and typical of much of eastern England.

This farm only ever received agrochemicals in the late 1940s. Since then it has been farmed organically. Approximately 6 ha of cereal crops are grown each year in rotation with grass and legume leys. Vegetable crops are grown on richer alluvial soil. The farm supports what is probably the largest population of Corn Buttercup left in Britain. Other uncommon arable plants include Shepherd's-needle and Night-flowering Catchfly. The owners became aware of the presence of Corn Buttercup when the feed merchant remarked on the quantity of seed in the winter Oat crop! The whole farm is designated as a County Wildlife Site.

No specific management is carried out for the arable flora, which thrives under the organic system. No herbicides are applied, and soil nutrient levels are maintained by the legumes in the crop rotation and the application of farmyard manure.

Creeping Thistle presents a problem occasionally, but otherwise the farmers are happy with their organic system and intend to continue with it.

A MANAGED FARM:
Kittyfield Farm at Melrose in Roxburghshire

Kittyfield Farm at Melrose, Roxburghshire occupies south-facing land by the River Tweed. Stony loams provide varying drainage from good to imperfect. The farm is owned and managed by Mr Luke Gaskell, a keen naturalist and chair of the Borders Farming and Wildlife Advisory Group (FWAG). Mr Gaskell, a member of the Botanical Society of the British Isles (BSBI), has found over 350 species of native flora at Kittyfield, including a number of uncommon arable plants. Cultivated land on the farm supports a rich arable flora, including the nationally scarce Tall Ramping-fumitory, Field Mouse-ear, Small-flowered Crane's-bill, Field Woundwort, Treacle-mustard. Great Lettuce, Musk Thistle and Henbit, Northern and Cut-leaved Dead-nettles also occur.

This arable land also supports farmland birds including Grey Partridge and Linnet. There are herb-rich pastures elsewhere on the farm. Although there is no official conservation designation at Kittyfield Farm, arable fields and pasture have been entered into the Central Borders Environmentally Sensitive Areas scheme.

Great Lettuce

ROTATIONAL MANAGEMENT:
Mwnt Arable Fields in Ceredigion, Wales

Summer on the Cliffs *by John Brett*

Mwnt Arable Fields lie on the Welsh coast just north of Cardigan (Aberteifi). The site is a little over 20ha in area, and comprises 3 blocks of fields near to north-facing cliffs overlooking Cardigan Bay. The soils are mainly clay, although near to the cliff edge soils derived from shales and mudstones also occur. The fields have 3 different owners, including the National Trust.

This has now been approved as a SSSI for its arable interest. The noted local botanist Arthur Chater has found a very rich community of arable plants in these fields, including the nationally scarce Small-flowered Catchfly (here at its best site in Ceredigion), a rich flora of fumitories, Sharp-leaved Fluellen, Weasel's-snout, Dwarf Spurge, Small-flowered Buttercup, Cornfield Knotgrass and Annual Knawel. On the evidence of *The New Atlas of the British and Irish Flora*, this latter species is one of the most rapidly-declining in Britain. Farmland birds, including Linnet and Skylark, use the fields, as do Chough, and occasionally Quail, which are rare summer visitors to Ceredigion.

The fields are currently under different forms of management: some are in arable cultivation, others are in set-aside, and some have been put down to grass. Traditionally, all manner of arable crops have been grown here, including Oats, Barley, Wheat, Beans, Potatoes, Carrots and Peas. These crops were grown as part of a rotation including fallow land or a grass or hay crop. The Countryside Council for Wales (CCW) is working with the owners to ensure that Mwnt arable fields are managed to safeguard their rare and important arable plant community.

 Royal Botanic Gardens, Kew During the summer of 2000, either side of the Broad Walk at the Royal Botanic Gardens, Kew was ablaze with a display of arable plants. These were interplanted with Wheat and included Corn Marigold, Corncockle and Common Poppy.

Since the great storm of 1987, there had been plans to restore the original vision of the landscape architect, William Nesfield, who laid out and planted the beds adjoining Decimus Burton's grand Broad Walk. This work was going to require replacing the large rectangular flower beds by small ones, so staff at Kew took the opportunity to make use of the site with an arable display to safeguard these particular plants. Kew has responsibility for Interrupted Brome, a grass first recorded in arable fields in 1849 but which is now regarded as extinct, having been last seen in Cambridge in 1972. Seed from the plant is stored in the Millennium Seed Bank based at Wakehurst Place, West Sussex, and Kew is researching the habitat requirements and history of the grass. As part of their British Biodiversity theme, Kew have sown Wheat with arable plants and put up interpretative boards. In addition, Cornflower and Interrupted Brome feature in their Threatened Plants Appeal. More information can be obtained from the Royal Botanic Gardens (see page 307).

Dwarf Spurge

Of 51 'target species' characteristic of arable land in the survey area, only 33 have been rediscovered. The only widely recorded target species are the fluellens and Dwarf Spurge. Prickly Poppy (on the Midvale Ridge sands) and Dense-flowered Fumitory (on the Chiltern chalk) still occur locally in good numbers, but the majority of species are found at very few sites, often in low numbers. For example, Red Hemp-nettle, historically widely recorded across the survey area, has been rediscovered at just two sites.

Since 1996, the Northmoor Trust, in collaboration with English Nature and (since 2000) Buckinghamshire County Council, has undertaken surveys to assess the current status and distribution of arable plants in Oxfordshire and Buckinghamshire. Areas known to have been historically rich have been targeted and surveyed thoroughly, field-by-field.

Thirty-seven farms (290 arable fields) on the limestone and calcareous sands of the Midvale Ridge Natural Area were surveyed during 1996–1998. Since 1999, surveys have concentrated on the chalk escarpment of the Chilterns Natural Area and a further 42 farms (692 arable fields) have been surveyed. As a rule, farmers and landowners have willingly granted permission to survey their land.

Soil type was the most important factor in determining arable plant diversity, with well-drained sandy or chalky soils supporting the rarest species. Spring-sown fields also tended to be more interesting than winter-sown ones.

Northmoor Trust has also researched changes in Oxfordshire's arable flora since the 1960s and made a comparison of the arable flora of organic and conventional farms.

Dense-flowered Fumitory

i **Plantlife, Wales** **In 2002 and 2003, Plantlife Wales, with help from the Countryside Council for Wales (CCW), devised a demonstration strip of arable flowers that they installed on the site of the Royal Welsh Agricultural Show held annually in Builth Wells. It stimulated a great deal of interest and people were encouraged to take a leaflet *Farming for Arable Flowers* which outlines things to do to manage land for arable flowers. More information can be obtained from the Plantlife Wales Officer (see page 307).**

Appendix 1
Nationally rare, nationally scarce and extinct arable plants in Britain

Common name	Scientific name	Status
Brome, Field	*Bromus arvensis*	**Unknown**
Brome, Interrupted	*Bromus interruptus*	**EXTINCT, ENDEMIC**
Brome, Rye	*Bromus secalinus*	**Unknown**
Bur-parsley, Lesser	*Caucalis platycarpos*	**EXTINCT**
Buttercup, Corn	*Ranunculus arvensis*	Nationally scarce
Catchfly, Small-flowered	*Silene gallica*	Nationally scarce
Chamomile, Corn	*Anthemis arvensis*	**Unknown**
Cleavers, Corn	*Galium tricornutum*	Nationally scarce
Cleavers, False	*Galium spurium*	**Unknown**
Corncockle	*Agrostemma githago*	**EXTINCT**
Cornflower	*Centaurea cyanus*	Nationally rare
Cornsalad, Broad-fruited	*Valerianella rimosa*	Nationally rare
Cudweed, Broad-leaved	*Filago pyramidata*	Nationally rare
Cudweed, Narrow-leaved	*Filago gallica*	**EXTINCT**
Cudweed, Red-tipped	*Filago lutescens*	Nationally rare
Fumitory, Dense-flowered	*Fumaria densiflora*	Nationally scarce
Fumitory, Few-flowered	*Fumaria vaillantii*	Nationally scarce
Fumitory, Fine-leaved	*Fumaria parviflora*	Nationally scarce
fumitory, Martin's Ramping-	*Fumaria reuteri*	Nationally rare
fumitory, Purple Ramping-	*Fumaria purpurea*	Nationally scarce, **ENDEMIC**
fumitory, Western Ramping-	*Fumaria occidentalis*	Nationally rare, **ENDEMIC**
Grass-poly	*Lythrum hyssopifolia*	Nationally scarce
Hedge-parsley, Spreading	*Torilis arvensis*	Nationally scarce
Hemp-nettle, Downy	*Galeopsis segetum*	**EXTINCT**
Hemp-nettle, Red	*Galeopsis angustifolia*	Nationally scarce
Lamb's Succory	*Arnoseris minima*	**EXTINCT**
Larkspur	*Colsolida ajacis*	**Unknown**
Pheasant's-eye	*Adonis annua*	Nationally rare
Quaking-grass, Lesser	*Briza minor*	Nationally scarce
Shepherd's-needle	*Scandix pecten-veneris*	Nationally scarce
Silky-bent, Dense	*Apera interrupta*	Nationally scarce
Silky-bent, Loose	*Apera spica-venti*	Nationally scarce
Speedwell, Fingered	*Veronica triphyllos*	Nationally rare
Spurge, Broad-leaved	*Euphorbia platyphyllos*	Nationally scarce
Thorow-wax	*Bupleurum rotundifolium*	**EXTINCT**
Vernal-grass, Annual	*Anthoxanthum aristatum*	**EXTINCT**
Vetch, Small-flowered	*Vicia parviflora*	Nationally scarce
Viper's-bugloss, Purple	*Echium plantagineum*	Nationally rare

Appendix 2
Archaeological records of some arable plants in Britain

Years before present	>5500 Pre-neolithic	5500-4000 Neolithic	4000-2500 Bronze Age	2500-2000 Iron Age	2000-1600 Roman	1600-1000 Saxon
Parsley-piert	●		●	●	●	●
Orache	●		●	●	●	●
Shepherd's-purse	●		●	●	●	
Cornflower	●				●	●
Fat-hen	●		●	●	●	●
Common Hemp-nettle	●		●	●	●	●
Knotgrass	●		●	●	●	●
Redshank	●		●	●	●	●
Corn Spurrey	●		●	●	●	●
Common Chickweed	●		●	●	●	●
Narrow-fruited Cornsalad	●		●	●	●	●
Rye Brome		●	●	●	●	●
Common Fumitory		●	●	●	●	●
Corn Poppy		●		●	●	●
Wild Radish		●	●	●	●	●
Charlock		●	●	●		
Field Penny-cress		●	●	●	●	●
Small Nettle		●	●	●	●	●
Scarlet Pimpernel		●	●	●	●	●
Corn Gromwell		●	●	●	●	●
Black-bindweed		●	●	●	●	●
Prickly Poppy		●	●	●	●	●
Corn Marigold		●	●	●	●	●
Broad-fruited Cornsalad			●	●	●	●
Shepherd's-needle			●	●		●
Night-flowering Catchfly			●	●	●	●
Darnel					●	●
Corn Buttercup					●	●
Corn Mayweed					●	●
Corncockle					●	●
Thorow-wax						●

Appendix 3
Flowering plants of arable land listed in the UK Biodiversity Action Plan

Status:
Rare – British Red Data Book
Scarce – Scarce Plants in Britain

	Status	BAP Listing
Broad-fruited Cornsalad	Rare	Priority List
Broad-leaved Cudweed	Rare	Priority List
Broad-leaved Spurge	Scarce	Species of Conservation Concern
Corn Buttercup	Scarce	Species of Conservation Concern
Corn Cleavers	Rare	Priority List
Field Gromwell		Species of Conservation Concern
Cornflower	Rare	Priority List
Fingered Speedwell	Rare	Species of Conservation Concern
Grass-poly	Rare	Species of Conservation Concern
Interrupted Brome	**EXTINCT**	Priority List
Martin's Ramping-fumitory	Rare	Species of Conservation Concern
Narrow-fruited Cornsalad		Species of Conservation Concern
Pheasant's-eye	Rare	Species of Conservation Concern
Purple Ramping-fumitory	Scarce	Priority List
Red Hemp-nettle	Scarce	Priority List
Red-tipped Cudweed	Rare	Priority List
Shepherd's-needle	Scarce	Priority List
Small-flowered Catchfly	Scarce	Priority List
Spreading Hedge-parsley	Scarce	Priority List
Western Ramping-fumitory	Rare	Priority List

Appendix 4
Flowering plants of occasionally cultivated land listed in the UK Biodiversity Action Plan

Status:
Rare – British Red Data Book
Scarce – Scarce Plants in Britain

Status	BAP Listing	
Cut-leaved Germander	Rare	Species of conservation concern
Downy Woundwort	Rare	Species of conservation concern
Ground-pine	Rare	Species of conservation concern
Hairy Mallow	Rare	Species of conservation concern
Perennial Knawel	Rare	Priority List
Perfoliate Penny-cress	Rare	Priority List
Sand Catchfly	Scarce	Species of conservation concern
Slender Bird's-foot-trefoil	Rare	Species of conservation concern
Small Alison	Rare	Species of conservation concern
Smooth Cat's-ear	Scarce	Species of conservation concern
Tower Mustard	Scarce	Priority List

Appendix 5
List of species mentioned in the text

PLANTS

Barren Brome	*Anisantha sterilis*
Black-grass	*Alopecurus myosuroides*
Black Bent	*Agrostis gigantea*
Bramble	*Rubus fruticosus* agg.
Bulbous Buttercup	*Ranunculus bulbosus*
Colt's-foot	*Tussilago farfara*
Cleavers	*Galium aparine*
Clustered Clover	*Trifolium glomeratum*
Creeping Buttercup	*Ranunculus repens*
Common Couch	*Elytrygia repens*
Common Field-speedwell	*Veronica persica*
Common Gromwell	*Lithospermum officinale*
Common Mallow	*Malva sylvestris*
Common Nettle	*Urtica dioica*
Common Hemp-nettle	*Galeopsis tetrahit*
Common Pignut	*Conopodium majus*
Common Ragwort	*Senecio jacobaea*
Common Stork's-bill	*Erodium cicutarium*
Corn Mint	*Mentha arvensis*
Corn/Blue Woodruff	*Asperula arvensis*
Cow parsley	*Anthriscus sylvestris*
Creeping Bent	*Agrostis stolonifera*
Creeping Thistle	*Cirsium arvense*
Crested Cow-wheat	*Melampyrum cristatum*
Daisy	*Bellis perennis*
Dwarf Mallow	*Malva neglecta*
Dwarf Mouse-ear	*Cerastium pumilum*
Eastern Larkspur	*Consolida orientalis*
Equal-leaved Knotgrass	*Polygonum arenastrum*
False Cleavers	*Galium spurium*
False Thorow-wax	*Bupleurum subovatum*
Fat-hen	*Chenopodium album*
Field Bindweed	*Convolvulus arvensis*
Field Forget-me-not	*Myostis arvensis*
Field Garlic	*Allium oleraceum*
Field Horsetail	*Equisetum arvense*
Field Mouse-ear	*Cerastium arvense*
Field Scabious	*Knautia arvensis*
Fine-leaved Sandwort	*Minuartia hybrida*
Flax Catchfly	*Silene linicola*
Flax Dodder	*Cuscuta epilinum*
Flixweed	*Decurania sophia*
Fool's Parsley	*Aethusa cynapium*
Forking Larkspur	*Consolida regalis*
Gold-of-pleasure	*Camelina sativa*
Great Lettuce	*Lactuca vicosa*
Groundsel	*Senecio vulgaris*
Hairy Tare	*Vicia hirsuta*
Hedge Bindweed	*Calystegia sepium*
Hoary Cinquefoil	*Potentilla argentea*
Italian Rye-grass	*Lolium multiflorum*
Ivy-leaved Speedwell	*Veronica hederifolia*
Knotgrass	*Polygonum aviculare*
Large-flowered Hemp-nettle	*Galeopsis speciosa*
Meadow Brome	*Bromus commutatus*
Meadow Buttercup	*Ranunculus acris*
Meadow Vetchling	*Lathyrus pratensis*
Musk Thistle	*Carduus nutans*
Northern Knotgrass	*Polygonum boreale*
Onion Couch	*Arrhenatherum elatius* var. *bulbosum*
Orache, Common	*Atriplex patula*
Pale Persicaria	*Persicaria lapathifolia*
Parsley-piert	*Aphanes arvensis*
Perennial Knawel	*Scleranthus perennis*
Perennial Rye-grass	*Lolium perenne*
Perennial Sow-thistle	*Sonchus arvensis*
Petty Spurge	*Euphorbia peplus*
Quaking-grass	*Briza media*
Red Bartsia	*Odontites vernus*
Red Dead-nettle	*Galeopsis angustifolia*
Redshank	*Persicaria maculosa*
Rosebay Willowherb	*Chamerion angustifolium*
Rough Marsh-mallow	*Althaea hirsuta*
Sand Catchfly	*Silene conica*
Scarlet Pimpernel	*Anagallis arvensis*
Scented Mayweed	*Matricaria recutita*
Scentless Mayweed	*Tripleurospermum inodorum*
Shepherd's Cress	*Teesdalia nudicaulis*
Shepherd's-purse	*Capsella bursa-pastoris*
Slender bird's-foot Trefoil	*Lotus angustissimus*
Small Cudweed	*Filago minima*
Small Medick	*Medicago minima*
Small Nettle	*Urtica urens*
Small-flowered Crane's-bill	*Geranium pussilum*
Smooth Cat's-ear	*Hypochaeris glabra*
Smooth Tare	*Vicia tetrasperma*
Soft-brome	*Bromus hordaceus*
Spear Thistle	*Cirsium vulgare*
Stinking/Scented Mayweed	*Matricaria recutita*
Stone Parsley	*Sison amomum*
Sweet Vernal-grass	*Anthoxanthum odoratum*
Toothed Medick	*Medicago polymorpha*
Treacle-mustard	*Erysimum cheiranthoides*
Tuberous Pea	*Lathyrus tuberosus*
Upright Hedge-parsley	*Torilis japonica*
Violet Horned Poppy	*Roemeria hybrida*
Viper's-bugloss	*Echium vulgare*
Wall Speedwell	*Veronica arvensis*
White Campion	*Silene latifolium*
White Ramping-fumitory	*Fumaria capreolata*
Wild Carrot	*Daucus carota*
Wild-oat	*Avena fatua*
Wild Radish	*Raphanus raphanistrum* ssp. *raphanistrum*

CROPS

Barley	*Hordeum vulgare*
Broad Bean	*Vicia faba*
Carrot	*Daucus carota* ssp. *sativus*
Flax/Linseed	*Linum usitatissimum*
Lentil	*Lens culinaris*
Oat	*Avena sativa*
Opium Poppy	*Papaver somniferum*
Garden Pea	*Pisum sativum*
Potato	*Solanum tuberosum*
Rye	*Secale cereale*
Sugar Beet	*Beta vulgaris* ssp. *vulgaris*
Wheat	*Triticum* spp.
Bread Wheat	*T. aestivum*
Emmer	*T. dicoccum*
Spelt	*T. spelta*
Einkorn	*T. monococcum*
Club Wheat	*T. aestivocompactum*

BIRDS

Chough	*Pyrrhocorax pyrrhocorax*
Cirl Bunting	*Emberiza cirlus*
Corn Bunting	*Emberiza calandra*
Corn Crake	*Crex crex*
Grey Partridge	*Perdix perdix*
Lapland Bunting	*Colcarius lapponicus*
Linnet	*Acanthis cannabina*
Quail	*Coturnix coturnix*
Reed Bunting	*Emberiza schoeniclus*
Skylark	*Alauda arvensis*
Snow Bunting	*Plectrophenax nivalis*
Stone-curlew	*Passer montanus*
Tree Sparrow	*Passer montanus*
Turtle Dove	*Streptopelia turtur*
Yellowhammer	*Emberiza citrinella*

ANIMALS

Brown Hare	*Lepus europaeus*
Rabbit	*Oryctolagus cuniculus*
Fairy Shrimp	*Chirocephalus* spp.

Appendix 6
Some commonly occurring arable plants

ENGLISH NAME	SCIENTIFIC NAME	PLANT/GRASS KEY PAGE No.	IDENTIFICATION NOTES
Bindweed, Field	Convolvulus arvensis	234	
Bugloss	Anchusa arvensis		Roughly hairy, 15–50 cm high with spear-shaped leaves and small blue flowers 5–6 mm wide
Charlock	Sinapsis arvensis		Typical cabbage, 30–80 cm high with unstalked toothed leaves and yellow flowers, 9–20 mm wide
Chickweed	Stellaria media		Low, often prostrate plant with oval, 3–20 mm long leaves and flowers 8–10 mm wide with 5 white, notched petals
Cleavers	Galium aparine	231	
Crane's-bill, Cut-leaved	Geranium dissectum		Typical crane's-bill with deeply-divided leaves
Fat-hen	Chenopodium album agg.		A very common, variable plant with soft whitish hairs giving it a grey downy appearance with spikes of inconspicuous green petalless flowers
Flixweed	Descurania sophia	230	
Forget-me-not	Myosotis arvensis		Familiar plant with small, 3–5 mm wide, grey-blue 5-petalled flowers
Goosefoot, Many-seeded	Chenopodium polyspermum		Like Fat-hen but hairless and with diamond-shaped irregularly toothed leaves
Hemp-nettle, Bifid	Galeopsis bifida		Like Common Hemp-nettle but with purple lower lip to flower
Hemp-nettle, Large-flowered	Galeopsis speciosa	241	
Knotgrass	Polygonum aviculare	233	
Knotgrass, Northern	Polygonum boreale	233	
Mayweed, Scented	Matricaria recutita	242	
Mayweed, Scentless	Tripleurospermum inodorum	242	
Mayweed, Stinking	Anthemis cotula	242	
Mint, Corn	Mentha arvensis	241	Possible to mistake for Field Woundwort
Mustard, White	Sinapsis alba		Similar to Charlock, but with deeply pinnately-lobed leaves
Nettle	Urtica urens		The familiar stinging nettle
Nightshade, Black	Solanum nigrum		5-petalled white flower with yellow centre reminiscent of a tomato flower; black poisonous berries
Parsley, Fool's	Aethusa cynapium	236	
Pimpernel, Scarlet	Anagallis arvensis	233	
Shepherd's-purse	Capsella bursa-pastoris	230	
Speedwell, Ivy-leaved	Veronica hederifolia		Typical speedwell with small pale lilac flowers and ivy-like leaves
Tare, Smooth	Vicia tetrasperma	240	
Brome, Barren	Anisantha sterilis	264	
Black-grass	Alopecurus myosuroides	264	
Oat, Wild	Avena fatua		

Fat-hen

Flixweed

Corn Mint

Shepherd's-purse

Selected Bibliography

ARABLE PLANTS

Britain's arable weeds. **P J Wilson**, 1992, *British Wildlife*, **3**: pp.149–161.

Fields of Vision – a Future for Britain's Arable Plants, Conference proceedings. Eds.: **P J Wilson and M King**, 2000 (English Nature and RSPB).

The Arable Weeds of Europe. **M Hanf**, 1983 (BASF UK Ltd).

The Ecology of Temperate Cereal Fields, 32nd Symposium of the British Ecological Society. Eds.: **L G Firbank, N Carter, J F Darbyshire and G R Potts**, 1991 (Blackwell Scientific Publications, Oxford).

Weeds and Aliens. **Sir E Salisbury**, 1961 (Collins New Naturalist).

Weeds of Farmland. **W E Brenchley**, 1920 (Longmans Green).

FABRIC: LANDSCAPE

An Illustrated History of the Countryside. **Oliver Rackham** (Weidenfield Illustrated, 1997)

Fields. **R and N Muir**, 1989 (Macmillan).

Fields in the English Landscape. **C Taylor** (Alan Sutton Publishing).

The History of the Countryside. **Oliver Rackham** (Cassell Illustrated, 1995)

The Land of Britain, its Use and Misuse. **L D Stamp**, 1962 (Longmans Green).

The Living Soil. **E B Barber**, 1953 (Faber & Faber).

FARMING

English Farming, Past and present. **Lord Ernle**, Ed.: **Sir A D Hall**, 5th Edn., 1936 (Longmans, Green and Co).

Organic Farming. **N Lampkin**, 1990 (Farming Press).

Suppression of Weeds by Fertilisers and Chemicals. **H C Long and W E Brenchley**, 1946 (Crosby and Lockwood).

The changing face of lowland farming and wildlife, part 2; 1945-1995. **C Stoate**, 1996, *British Wildlife*, **7**: pp.162–172.

Post-war changes in arable farming and biodiversity in Great Britain. **R A Robinson and W J Sutherland**, 2002, *Journal of Applied Ecology*, **39**: pp.157–176.

The Partridge, Pesticides, Predation and Conservation. **G R Potts**, 1986 (Collins).

KEY GAME CONSERVANCY TRUST PAPERS ON FARMLAND WILDLIFE

Studies on the cereal ecosystem. **G R Potts and G P Vickerman**, 1974, *Advances in Ecological Research*, **8**: pp.107–97.

Reduced pesticide inputs on cereal field margins: the effects on butterfly abundance. **J W Dover, N W Sotherton and K Gobbett**, 1990, *Ecological Entomology*, **15**: pp.17–24.

Twenty years of monitoring invertebrates and weeds in cereal fields in Sussex. **N J Aebischer**, 1991. In: *The Ecology of Temperate Cereal Fields*. Eds.: **L G Firbank, N Carter, J F Darbyshire and G R Potts**. pp. 305–22 (Blackwell Scientific Publications, Oxford).

Where the Birds Sing. The Allerton Project: 10 years of conservation on farmland. **C Stoate and A Leake**, 2002, The Game Conservancy Trust, Fordingbridge, Hampshire.

HISTORY

New Studies in Archaeology – Prehistoric Farming in Europe. **G Barker**, 1985 (Cambridge University Press).

A Regional History of England. Series edited by **Barry Cunliffe** and **David Hey**, 1993 (Longmans).

The History of the British Flora. **Sir H Godwin**, 1956 (Cambridge University Press).

PLANTS

Britain's Rare Flowers. **Peter Marren**, 1999 (Poyser Natural History).

Flora Britannica. **Richard Mabey**, 1996 (Sinclair Stevenson).

Lebensraum Acker. **H Hofmeister** and **E Garve**, 1986 (Paul Parey).

New Flora of the British Isles, 2nd Edn. **C Stace**, 1997 (Cambridge University Press).

Red Data Book 1: Vascular plants, 3rd Edn. Ed.: **M Wigginton**, 1999 (JNCC).

Scarce Plants in Britain. Eds.: **N Stewart, D A Pearman and C D Preston**, 1994 (JNCC).

Seeds. **C C Baskin** and **J M Baskin**, 2001 (Academic Press).

The Englishman's Flora. **Geoffrey Grigson**, 1958 (Paladin).

The Illustrated Book of Food Plants, 2nd Edn., 1985 (Oxford University Press).

The New Atlas of the British and Irish Flora. **C D Preston, D A Pearman** and **T D Dines**, 2002 (Oxford University Press).

PLANT IDENTIFICATION

A Field Guide to the Crops of Britain and Europe. **G M De Rougemont**, 1989 (Collins).

Flora of the British Isles, 3rd Edn. **A R Clapham, T G Tutin** and **D M Moore**, 1987 (Cambridge University Press).

The Wild Flower Key. **F Rose**, 1981 (Warne).

Wild Flowers of Britain and Northern Europe. **R and A Fitter, M Blamey**, 1987 (Collins).

Grasses. **C E Hubbard**, 1954 (Penguin Books).

British Mosses and Liverworts. **E V Watson**, 1981 (Cambridge University Press).

Art and photographic credits

All black and white illustrations by Rick Havely unless stated.

Useful names and addresses

ADAS
Head Office, Woodthorne, Wergs Road, Wolverhampton WV6 8TQ

Tel: 01902 754190
www.adas.co.uk

Provides consultancy and undertakes research into agricultural issues.

Biodiversity Action Plans

The Action Plans on the website have superseded the printed Action Plans and include revised targets. Site gives names of Lead Partners and Contact Points.

www.jncc.org or **www.ukbap.org**

The Cereal Field Margins HAP and the individual arable plant SAPs can be found there. The arable plants group which focuses on issues affecting these plants has a web site:

www.arableplants.org.uk

Botanical Society of the British Isles (BSBI)
C/o Botany Department, Natural History Museum, Cromwell Road, London SW7 5BD

Pete Selby, Volunteers' Officer,
Tel: 02380 644368
www.bsbi.org.uk

A charity whose members comprise professional and amateur botanists dedicated to the study of vascular plants and stoneworts in the UK. Has a national system of Vice County Recorders who keep records of all local sightings. Local BSBI members have established groups who could be willing to help in counties such as Cornwall, Somerset, Sussex and Oxfordshire. The local Recorder should have details.

Butterfly Conservation
Manor Yard, East Lulworth, near Wareham, Dorset BH20 5QP

Tel: 01929 400209
www.butterfly-conservation.org

Charity concerned with the conservation of butterflies and moths and their habitats.

Council for the Protection of Rural England (CPRE)
National Office: 128 Southwark Street, London SE1 0SW

Tel: 020 7981 2800
www.cpre.org.uk

Registered charity which exists to promote the beauty and diversity of rural England by encouraging the sustainable use of land.

Country Land and Business Association:
16 Belgrave Square, London SW1X 8PQ

Tel: 020 7235 0511
www.cla.org.uk

Represents those who live and work in the countryside.

Countryside Agency
John Dower House, Crescent Place, Cheltenham GL50 3RA

Tel: 01242 521381
www.countryside.gov.uk

Contact for National Parks, Areas of Outstanding Natural Beauty and a wide range of countryside matters.

Crop Protection Association:
4 Lincoln Court, Lincoln Road, Peterborough PE1 2RP

Tel: 01733 349225
www.cropprotection.org.uk

Trade body representing a range of companies.

Countryside Council for Wales (CCW)
Plas Penrhos, Ffordd Penrhos, Bangor, Gwynedd LL57 2LQ, Wales

Tel: 01248 385500
www.ccw.gov.uk

Contact for all matters concerning countryside conservation including Sites of Special Scientific Interest in Wales and the Tir Gofal Scheme.

Department of Agriculture and Rural Development Northern Ireland
Countryside Management Division Dundonald House, Upper Newtownards Road, Belfast BT4 3SB

Tel: 02890 520100
www.dardni.gov.uk

Contact for information on Environmental Land Management Schemes in Northern Ireland.

Department for Environment, Food and Rural Affairs (Defra)
Conservation Management Division
Nobel House, 17 Smith Square, London SW1P 3JR

Tel: 020 7238 6000
www.defra.gov.uk

Contact for information on Environmental Land Management Schemes in England. Also co-ordinates implementation of the UK Biodiversity Action Plan and

acts as Lead Partner for Cereal Field Margins Habitat Action Plan and as Contact Point for 10 species of arable plants under the UK Biodiversity Action Plan.

Wildlife and Conservation Division

Temple Quay House, 2 The Square, Temple Quay, Bristol BS1 6BB

Tel: 0117 372 8974;
E-mail: biodiversity.defra@gtnet.gov.uk

Elm Farm Research Centre

Hamstead Marshall, Newbury RG20 0HR

Tel: 01488 658298
www.efrc.com/

Educational charity founded in 1980 provide information and advice on organic farming.

English Nature

Northminster House, Peterborough PE1 1UA

Tel: 01733 455000
www.english-nature.org.uk

Government advisory body for all matters concerning nature conservation in England, Sites of Special Scientific Interest and the Wildlife Enhancement Scheme. Lead agency for two species of arable plants under the UK Biodiversity Action Plan.

Environment Agency

Head office: Rivers House, Waterside Drive, Aztec West, Almondsbury, Bristol BS12 4UD

Tel: 01452 624400
www.environment-agency.gov.uk

Head office in Bristol, 7 Regional offices and Environment Agency, Wales plus 26 Area offices. Enforces environmental pollution legislation

Environment and Heritage Service Northern Ireland

Commonwealth House, 35 Castle Street, Belfast BT1 1GU, Northern Ireland.

Tel: 029 9025 1477
www.nics.gove.uk/ehs/

Contact for matters relating to the conservation of the natural and built heritage including responsibility for Areas of Special Scientific Interest (ASSIs)

Farming and Wildlife Advisory Group (FWAG)

National Agriculture Centre, Stoneleigh, Kenilworth, Warks CV8 2RX
Tel: 02476 696760
FWAG Cymru: Tel: 01341 421456;
FWAG Scotland Tel: 01314 724080/1
www.fwag.org.uk

Has over 90 Farm Conservation Advisers working locally to provide advice to farmers and landowners taking a whole farm approach.

Flora locale

www.naturebureau.co.uk/pages/floraloc/floraloc.htm
email: floralocale@hotmail.com

Established in 1997 to promote the conservation of native wild plants and to provide advice.

Forum for the Future

227a City Road, London EC1V 1JT

Tel: 020 7251 6070
www.forumforthefuture.org.uk

Working to devise and implement sustainable solutions including project on sustainable agriculture.

Game Conservancy Trust

Burgate Manor, Fordingbridge, Hants SP6 1EF

Tel: 01425 651021
www.gct.org.uk

Registered charity providing advice on game management and sustainable farming.

Institute for European Environmental Policy (IEEP)

Dean Brambley House, 52 Horseferry Rd, London SW1P 2AG

www.ieep.org.uk

Core policy area focused on agriculture and rural development paying particular attention to agriculture and European Union policy.

International Institute for Environment and Development (IIED)

3 Endsleigh Street, London WC1H 0DD

Tel: 0207 388 2117
www.iied.org.uk

Independent non-profit organisation promoting sustainable development. Includes a biodiversity and livelihoods group.

Landlife

National Wildflower Centre, Court Hey Park, Liverpool L16 3NA

Tel: 0151 737 1819
www.landlife.org.uk

Charity working in mainly urban areas to bring people and wildlife together.

National Assembly for Wales Agriculture Department

Crown Buildings, Cathays Park, Cardiff CF1 3NQ

Tel: 029 20825111
www.wales.gov.uk/subiagriculture

Contact for information on Environmentally Sensitive Areas in Wales.

National Biodiversity Network

Secretariat: The Kiln, Mather Road, Newark
NG24 1WT

Tel: 01636 670090
www.nbn.org.uk

*Partnership committed to making information about
UK wildlife available via the web.*

National Farmers Union

Agriculture House, 164 Shaftesbury Avenue,
London W1ZH 8HZ

Tel: 020 7331 7200
www.nfu.org.uk

*Represents farmers and growers in England and
Wales. Organisation comprises London HQ, 2 Welsh
offices, 7 regional and over 300 branch offices.*

National Trust

33 Sheep Street, Cirencester, Gloucestershire
GL7 1RQ

Tel: 01285 651818
www.nationaltrust.org.uk

*Charitable body concerned with the conservation of
places of historic interest and natural beauty in
England, Wales and Northern Ireland.*

National Trust for Scotland

28 Charlotte Square, Edinburgh EH2 4ET

Tel: 0131 243 9300
Fax: 0131 243 9301
www:nationaltrust.org.uk

Northmoor Trust

Little Wittenham, Abingdon, Oxon OX14 4RA

Tel: 01865 407792
www.northmoortrust.org.uk/ecology

*Promotes conservation through good practice,
education and land science. Has carried out numerous
surveys of farms for arable plants.*

Plantlife

21 Elizabeth Street, London SW1W 9RP
Tel: 020 7808 0100
www.plantlife.org.uk

Wales
Plantlife Wales Officer, c/o Countryside Council
for Wales, Maes y Ffynnon, Ffordd Penrhos,
Bangor, LL57 2LQ.

*Charitable body concerned with the conservation of
wild plants and their habitats. Lead partner for BAP
listed arable plants and co-ordinates the Arable
Plants Group*

Royal Botanic Gardens

*Gardens open to he public and centres for botanical
science.*

Edinburgh
20a Inverleith Row, Edinburgh EH3 5LR

Tel: 0131 552 7171
www/rbge.org.uk

Kew
Royal Botanic Gardens, Kew, Richmond, Surrey
TW9 3AB
020 8332 5000
www.rbgkew.org.uk

West Sussex
Wakehurst Place, Ardingly, Nr Haywards Heath,
West Sussex RH17 6TN

Tel: 01444 894066

Wales
National Botanic Garden of Wales, Llanarthne,
Carmarthenshire SA32 8HG

Tel: 01558 667150
www:gardenof wales.org.uk

Royal Society for the Protection of Birds

The Lodge, Sandy, Bedfordshire SG19 2DL

Tel: 01767 680551
www.rspb.org.uk

*Charitable organisation concerned with the
conservation of wild birds and their habitats.*

Rural Development Service/Defra

*Provides an advisory service. Manages the England
Rural Development Programme on behalf of Defra
and gives advice on Agri-Environment agreements.*
**www.defra.gov.uk/corporate/rds/
default.asp**

Bristol
Block 3, Government Buildings, Burghill Rd,
Westbury on Trym, Bristol BS10 6NJ

Tel: 0117 959 1000

Cambridge
Block B, Government Buildings, Brooklands
Ave, Cambridge CB2 2DR

Tel: 01223 462727

Crewe
Electra Way, Crewe Business Park, Crewe,
Cheshire CW1 6GJ

Tel: 01270 754000

Leeds
Government Buildings, Otley Road,
Lawnswood, Leeds LS16 5QT

Tel: 0113 230 3750

Newcastle
Government Buildings, Kenton Bar, Newcastle-upon-Tyne NE5 3EW
Tel: 0191 214 1800

Nottingham
Block 7, Government Buildings, Chalfont Drive, Nottingham NG8 3SN
Tel: 0115 929 1191

Reading
Government Buildings, Coley Park, Reading RG1 6DT
Tel: 0118 958 1222

Worcester
Block C, Government Buildings, Whittington Road, Worcester WR5 2LQ
Tel: 01905 763355

Scottish Executive Rural Affairs Department
Pentland House, 47 Robb's Loan, Edinburgh EH14 1TY
Tel: 0131 556 8400
www.scotland.gov.uk
Contact for information on Environmental Land Management Schemes in Scotland (ESAs and the Rural Stewardship Scheme)

Scottish Environmental Protection Agency (SEPA)
Erskine Court, Castle Business Park, Stirling FK9 TR
Tel: 01786 457700
www.sepa.org.uk
Aims to implement integrated environmental protection system for Scotland.

Scottish Natural Heritage
2-3 Anderson Place, Edinburgh EG9 2AS
Tel: 0131 446 2277
www.snh.org.uk
Contact for all matters concerning countryside conservation and Sites of Special Scientific Interest in Scotland.

The Soil Association
40-56 Victoria St, Bristol BS1 6BY
Tel: 0117 929 0661
www.soilassocation.org.uk
Formed in 1946 this organisation concentrates on organic food and farming.

Universities and research organisations:
An alphabetical list can be found on:
www.scit.wlv.ac.uk/ukinfo

Aberdeen, Aston, Bradford, Brighton, Bristol, Brunel, Cambridge, Durham, East Anglia, Edinburgh, Essex, Greenwich, Hertfordshire, Keele, Kent, Kingston, Lancaster, Leeds, Leicester, Lincoln, Liverpool, London – Imperial College including Wye College, Queen Mary Westfield, and University College, London; Loughborough, Manchester, Newcastle, Nottingham, the Open University, Oxford, Reading, Stirling, Sussex, Ulster, University of the West of England, Bristol; Wales: Aberystwyth and Bangor; and York

Institutes:
CABI Bioscience, Elm Farm Research Centre, Harper Adams university College,
Institute of Arable Crop Research at Brooms Barn, Long Ashton and Rothamstead; Institute of Grassland and Environmental Research at Aberystwyth and Okehampton; Institute of Plant Science Research, Macaulay Land use Research Institute, and NERC CEH at Banchory, Bangor, Edinburgh, Furzebrook, Merlewood – joining Lancaster University, Monk's Wood, Oxford, Wallingford and Windermere; Royal Agricultural college, Cirencester, Scottish Agricultural College, Scottish Crop Research Institute, Writtle College

The Wildlife Trusts
UK office, The Kiln, Waterside, Mather Road, Newark NG24 1WT
Tel: 01636 677711
www.wildlifetrusts.org
Voluntary conservation organisation concerned with the conservation of wildlife throughout the UK. Contact for information on the 48 County Wildlife Trusts.

Acknowledgements

Many people have contributed to the production of this book and our sincere thanks are due to them all. It is our intention that everyone who has contributed to the book is named in it, but if we have missed anyone inadvertently we can only apologise. Despite the contributions of others, we hold full responsibility for any errors that may have crept in, and for any omissions which we may have made.

Photographs: The photographs are one of the key features of this book and would not have been possible without the help and co-operation of all the photographers whose work is featured. The names of all the photographers who took each photograph is shown in the credits starting on page 304 but particular thanks are due to Dr. Chris Gibson of English Nature for the quantity and quality of photographs supplied, to Bob Gibbons of Natural Image and Andrew Gagg (Agency?) and to Richard Lansdown for his fumitory images. Thanks also to Stan Dumican; Christine Saw of the Butser Anceint Farm; Ian Bennallick, Environmental Records Centre for Cornwall and the Isles of Scilly; Richard Scott, Landlife; Simon Murphy of London's Transport Museum; Lindsey Butterfield and Simon Ford of the National Trust; Andy Jackson, Royal Botanic Gardens, Kew; Leigh Lock RSPB Southwest; Liz McDonnell, James Phillips and Mark Stevenson, RDS; Dr Alan Knapp and members of the Sussex Botanical Group and Gina Kittle of the Weald and Downland Museum, Singleton, West Sussex. who went to particular trouble in locating and supplying photographs.

Text: Many people have helped us by supplying information, proof reading and editing the text including Dr Chris Gibson, Phil Grice, Anna Hope, Rebecca Isted, Dr Brian Johnson, Paul Lacey, Simon Leach, Karen Mitchell, Haydn Pearson, Ron Porley, Dr Jill Sutcliffe, James Trueman, Val Wheeler (English Nature staff); Huw Lewis and Helen Evans (CCW); John Henderson and Donald Bailey (Scottish Executive Rural Affairs Department), Liz McDonnell and Dave Smallshire (Defra), Pat Neylon (FWAG); Steve Gregory and Susanna Kay (Northmoor Trust). Particular thanks are due to Liz McDonnell and Ro Fitzgerald for their detailed input to the book and to Rose Murphy of the Botanical Cornwall Group for her knowledge, enthusiasm and invaluable contribution concerning the fumitories. In addition, the whole text benefitted from the editing and design skills of Andy and Gill Swash and Rob Still of WILD*Guides*.

Illustrations: A debt of gratitude is owed to John Davis for his incredible speed at turning round the commissioned coloured artwork and to Rick Havely for dealing with the ever increasing list of illustrations within a non-moving deadline. Thanks also to Ani Overton and Marian Reed for supplying figures.

Maps: We would also like to thank Dr Chris Preston and Henry Arnold at the Centre for Ecology and Hydrology, Monk's Wood for supplying the map data with such efficiency produced from the data collected by the BSBI and published in *The New Atlas of the British and Irish Flora* edited by Dr Crhis Preston, David Opearman and Dr Trevor Dines (OUP, 2002).

Case studies: Thanks to the contributors to the case studies and other 'boxed' sections: the Thoroughgood family – Bushy Ley Farm; Christopher Clarke – Cridmore Farm; David Northcote-Wright – Fivehead Arable Fields; Somerset Wildlife Trust; Thomas Eggers – Germany; Rona Charles, North Yorkshire Moors National Park – Silpho; Simon Ford, National Trust – West Pentire; Mr Maynard – Whittlesford and Thriplow Hummocky fields, David Strange – Worth and Compact Farm; Luke Gaskell – Kittyfield Farm; Andrew Jones (CCW) and Philip Mould, Chairman of Plantlife International – Mwnt Arable Fields; Steve Gregory – Northmoor Trust; and Peter Thompson, Game Conservancy Trust.

Index